RADIATION
and LIFE
Second Edition

Pergamon Titles of Related Interest

Bentel TREATMENT PLANNING AND DOSE CALCULATION IN
RADIATION ONCOLOGY, Third Edition
Cember INTRODUCTION TO HEALTH PHYSICS, Second Edition
Eisenberg RADIATION PROTECTION, 2 Volume Set
Gollnick EXPERIMENTAL RADIOLOGICAL HEALTH PHYSICS
Graham GRENZ RAYS
Kase/Nelson CONCEPTS OF RADIATION DOSIMETRY

Related Journals*

ANNALS OF ICRP
HEALTH PHYSICS
INTERNATIONAL JOURNAL OF APPLIED RADIATION AND
ISOTYPES
INTERNATIONAL JOURNAL OF RADIATION
ONCOLOGY/BIOLOGY/PHYSICS

*Free specimen copies available upon request.

RADIATION
and LIFE
Second Edition

ERIC J. HALL
College of Physicians & Surgeons of
Columbia University

PERGAMON PRESS
New York   Oxford   Toronto   Sydney   Paris   Frankfurt

Pergamon Press Offices:

U.S.A.   Pergamon Press Inc., Maxwell House, Fairview Park,
              Elmsford, New York 10523, U.S.A.

U.K.   Pergamon Press Ltd., Headington Hill Hall,
              Oxford OX3 0BW, England

CANADA   Pergamon Press Canada Ltd., Suite 104, 150 Consumers Road,
              Willowdale, Ontario M2J 1P9, Canada

AUSTRALIA   Pergamon Press (Aust.) Pty. Ltd., P.O. Box 544,
              Potts Point, NSW 2011, Australia

FRANCE   Pergamon Press SARL, 24 rue des Ecoles,
              75240 Paris, Cedex 05, France

FEDERAL REPUBLIC   Pergamon Press GmbH, Hammerweg 6,
OF GERMANY   D-6242 Kronberg-Taunus, Federal Republic of Germany

Copyright © 1984 Pergamon Press Inc.

Library of Congress Cataloging in Publication Data

Hall, Eric J.
    Radiation and life

    Includes bibliographical references and index.
    1. Radiation--Physiological effect. 2. Radiation--
Safety measures. 3. Radiology, Medical. 4. Atomic
power. I. Title.
RA569.H33   1984   612'.01448   83-25708
ISBN 0-08-028819-7

Printed in Great Britain by A. Wheaton & Co. Ltd., Exeter

Contents

Preface

The use of radiation has become an integral part of modern life. From the x-ray picture of a broken limb to the treatment of cancer, the medical applications of radiation are now accepted as commonplace, and x-ray facilities are available in every community hospital. If the lights are not to go out in the big cities, it is difficult to imagine how sufficient electricity can be generated in the next decade without the help of nuclear reactors.

Living creatures have always received some radiation from natural sources, such as radioactivity in the ground and in food, as well as cosmic rays from outer space. Man-made radiations, including medical and industrial uses of x-rays and atomic energy, result in an additional exposure above and beyond that which we receive naturally. As a result of modern technology, astronauts have journeyed to the moon, and millions of people each year fly in commercial jet-liners; in both cases an extra exposure to cosmic rays is involved.

Radiation in medicine and in industry undoubtedly results in tremendous benefits to society; we know this for sure. An excessive dose can result in disastrous consequences to our health and well-being; we are sure of this, too. Radiation can cause disease as well as cure it. Even the invaluable x-ray picture, incautiously and excessively used, while not harming our own bodies in any perceptible way, can alter our genetic make-up and lead to deleterious effects in our descendants. The area of doubt and uncertainty, where we are not at all sure, includes the trade-off between risk and benefit. What dose of radiation is "excessive"? Is any dose safe? How much risk are we prepared to accept as the price for how many benefits?

No attempt will be made to discuss nuclear weapons used in war; a discussion of this topic is outside the purview of this book. If the day ever dawns when major nations, in all-out conflict, launch ballistic missiles armed with nuclear warheads, then civilization—as we know it—will end. In such a circumstance the radiation effects become virtually irrelevant because they are dwarfed into insignificance by the destructive force of the blast and heat. This book is devoted exclusively to the peaceful uses of the atom and the radiation exposure to man that results.

My mission is twofold: first, to describe, in terms readily understood by the nonspecialist, what radiation is and how it affects living things; second, to review the various sources of radiation to which man is exposed as a result of modern technology, and to view them in perspective. In particular, attention will be focused on a comparison of medical and industrial uses of radiation. It has been fashionable to scream loudly in protest when a utility company proposed to build a nuclear reactor to generate electricity. The chances are that the same protestors would submit meekly to a series of dental x-rays every 6 months. The question is, does this attitude make good sense?

The fear of radiation produces the most irrational response in people who otherwise are calm and intelligent. To some extent this is understandable, since radiation is invisible, and produces subtle biological changes that usually are not apparent for many years. In addition, radiation has traditionally received a bad press. It is indelibly associated in the minds of most people with bombs, fall-out, death and destruction.

Knowledge is the best antidote to fear and suspicion. The more people know about radiation, the benefits it confers and the risks it entails, the better able they will be to play their part as good citizens in deciding the role of radiation in our society. For ultimately it is the people, through their elected representatives, who will decide how much electricity is to be generated by nuclear power, and how careful will be the controls imposed in the medical use of x-rays. If this book helps, in some small way, to educate the public to make a wise and rational choice, the author will feel amply rewarded.

E.J.H.
New York
August 1975

Preface to the Second Edition

This second edition is produced in response to many requests from individuals who found the first edition of some use as a simple introduction to the uses of radiation but who felt that an update would be timely. Radiation has been much in the news since the first edition appeared, and yet the overall conclusions, based on science and reason, have altered little. The incident at Three Mile Island sparked great public interest in the effects of radiation from nuclear power reactors, but this interest did not spill over to the much bigger problem of medical x-rays, which involve far higher doses. Organized protests against nuclear power have been very effective in the United States, to the extent that the nuclear industry has been brought virtually to a halt. As a consequence, in the 1970s, when nuclear power was needed most because of escalating oil prices, it was simply not available in quantity. The effect of the protestors, therefore, was substantial on the economy as a whole, far transcending the nuclear industry. These events were a direct consequence of the American brand of democracy, which invites public participation in major decisions—far more so than is the case in other Western countries—thus providing a forum for the protestors, both sincere and insincere, with legitimate objections and otherwise.

The scene in Western Europe is mixed. In some countries, such as West Germany, the nuclear program is blocked as effectively as in the United States. The effects of this are apparent around the Ruhr Valley where the air is thick with smoke and soot from giant power stations forced to burn low-grade coal to meet the bustling economy's voracious appetite for energy. In other countries of Europe, including France and Great Britain, the nuclear power program is quietly but steadily accelerating. Many individuals in the regulatory agencies perceive that the battle to convince the public of the need and inherent safety of nuclear power has been won. This may represent premature optimism.

The decision was made during the preparation of the first edition of this book to exclude the consideration of nuclear weapons. This decision is maintained in this revision, except insofar as the proliferation of nuclear reactors for the generation of power may contribute directly or indirectly to the proliferation of nuclear weapons by making a supply of fissionable material more readily available. Otherwise it makes little sense to discuss nuclear weapons in the same context as medical x-rays and nuclear power production. If unlimited nuclear weapons were used in an all-out war, society as we know it would presumably come to an end. The radiation aspect would be dwarfed into insignificance by the devastation produced by blast and heat, as was the case at Hiroshima and Nagasaki. In the A-bomb attacks on Japan, 100,000 people were killed on impact, while the late effects of radiation comprise 400 cases of cancer and leukemia in survivors of the blast, compared with the 16,000 expected cases of cancer from natural causes.

In this second edition, the chapters concerning the biological effects of low doses of radiation have been updated, revised and considerably expanded to include the latest data and ideas on carcinogenesis and the genetic effects of radiation. Many new pictures and diagrams have been added to help illustrate the basic concepts. Scientists know more about the cancer-causing potential of ionizing radiations than about any other environmental carcinogen. There is a wealth of epidemiological information about irradiated human populations, spanning from the survivors of the atom bombs dropped at Hiroshima and Nagasaki to patients exposed to medical x-rays, as well as an extensive literature about radiation effects on experimental animals and cells. Yet the cancer risks from low doses of radiation remain a subject of bitter scientific dispute. There are two fundamental difficulties. First, it is not possible to obtain clear direct evidence of cancer caused by radiation at the very low levels characteristic of current medical practice or emanating from nuclear power reactors. The additional incidence of radiation-induced cancer is so small compared with the natural or spontaneous levels in the human that the sample sizes that would need to be studied to obtain statistically convincing information are impractically large. Consequently, the only practical approach is to observe much smaller groups of people exposed to far higher levels of radiation: patients treated with medical x-rays or the Japanese atom bomb survivors, for example. But then scientists come up against the second fundamental difficulty: they do not really know how to extrapolate from high doses to low doses. The simplest solution is to assume that cancer risk is proportional to dose, i.e., that if 1 Gray of x-rays produces, say, 500 cases of cancer, then a tenth of the dose (0.1 Gray) would produce 50 cases. However, there are good reasons to believe that this assumption may be naive and not generally applicable to carcinogenesis, though it may be more reasonable for the genetic effects of radiation. This dilemma is at the center of the bitter controversy over the effects of low level radiation, and in this book an attempt is made to explain the problems in simple, nontechnical terms.

E.J.H.
New York
July 1983

Acknowledgments

This book covers a broad range of subjects, from medicine to power production, from biology to space flight. No one person could possibly speak with authority or from personal experience on such diverse topics; certainly the author does not claim to do so.

I sought help from many people. Some were friends and colleagues of long standing; others were solicited because of their acknowledged expertise in a particular field. Without exception, they all responded generously. It is my pleasure to record the debt of gratitude that I owe to them for counsel and advice, for the critical review of the text, and for providing additional material. They include:

Dr. Victor Bond	Dr. Raymond Oliver
Dr. Stanley Curtis	Dr. Vance Sailor
Dr. H.V. Davenall	Dr. William Seaman
Dr. Philip Johnson	Dr. Herman Suit
Dr. Henry Kaplan	Dr. Edith H. Quimby
Mr. Phil Lorio	Dr. Irving Lerch

Many individuals and companies provided photographs and diagrams which are acknowledged in the text.

This book is intended for the nonspecialist, for the interested "man in the street." To this end, my wife, Bernice, who has no training as a scientist, read the manuscript time and time again to censure any unduly technical terms and veto subjects when they became too complex. If the reader finds this book easy to follow it is due in no small measure to her efforts. My secretary, Mrs. Jeanne Kramer, worked tirelessly to type and check the many revisions of the manuscript; I would not miss this opportunity to applaud her skill and thank her for her loyalty.

My own research is supported by the United States Department of Energy and by the National Cancer Institute.

RADIATION
and LIFE
Second Edition

1
Our Radiation Heritage

THE RADIATION ENVIRONMENT

Life on earth has developed with an ever-present background of radiation. It is not something new, invented by the wit of man; radiation has always been there. An issue often debated is whether life has evolved in spite of the potential deleterious effects of radiation—the winner in a constant battle—or whether the ability of radiation to cause mutations has been a vital factor in the continual upward evolution of biological species. No one is sure at present which is the case, and it is probable that the answer will never be known with any certainty.

What is new, what is man-made, is the extra radiation to which we are subjected from medical x-rays in the hospital or dentist's office, from journeys in high-flying jet aircraft, from the fallout of nuclear weapons testing, and from nuclear reactors built to generate electrical power. There can be no denying that a man-made component of radiation is being added continuously to the background level which we receive naturally. This is a cause of great concern to the public, and must ultimately implicate the whole of society because of the critical choices and issues involved.

Many of the pollutants which we face as a by-product of this technological age are new and unique in the sense that no creature, human or otherwise, has ever had to contend with them before. For instance, many chemicals used for food additives or pesticides, much of the smoke and products of burning coal and oil, did not exist on earth in significant quantities until man made them. They are a totally new hazard faced by mankind, and of his own making. No animal in its natural habitat has ever continuously inhaled smoke and the products of combustion. This is a new experience reserved for the factory workers of the nineteenth century and every city dweller of the twentieth.

Radiation is not like this. It has always been present. What we are doing nowadays is adding to the existing background an extra dose of radiation from man-made devices. There is essentially no difference in kind between natural and man-made radiations. Radiation is essentially different from other forms of pollution and this difference could turn out, in the final analysis, to be vital. Biological systems have a remarkable capacity to adapt to situations to which they are gradually exposed for a long period of time.

The initial confrontation with a new toxic agent may be devastating, but the effect is diluted as the organism adapts and evolves. Life in all forms on earth has evolved from the dawn of time against a continual background of radiation, and there is every reason to believe that living things are well able to cope, provided the levels are not too high.

THE "NATURAL LIFE"

There is so little that is "natural" in any aspect of our daily lives. Those who advocate a return to the natural have either not thought about it, or have closed their eyes to the realities involved. It is "natural" for one in ten children to die at birth, and for a similar number of mothers to lose their lives during this most natural of all functions. A considerable amount of medical intervention is necessary to achieve the low mortality rates which we accept today as par for the course.

It is "natural" that a whole generation of children should be wiped out periodically by diptheria or smallpox, and that the weak of each generation should be debilitated and die an early death as a result of the hacking cough and gradual asphyxiation characteristic of tuberculosis. The methods we use to ensure the health of our children are artificial and man-made.

Our homes today are temperature-controlled incubators; warmed artificially in winter and cooled in summer. We are transported about at high speed in motorcars or airplanes, fueled with oil, and eat an abundance of cheap food because nature's shortcomings are supplemented with fertilizers and nature's excesses curbed with pesticides. The city dweller in a northern climate, up to his knees in snow during midwinter, can eat asparagus and green salad followed by fresh strawberries. In order to provide Americans with the variety, prices, quantities, consistencies and qualities of food they demand, food companies over the years turned to an array of food additives, preservatives, flavor enhancers, colorings, and other ingredients. These products make food look and taste better and have made possible the "mass merchandising" of food that today's consumers accept as a way of life. There is little that is natural about our life today.

We may, from time to time, make a token rebellion against this state of affairs. We may long for the hand-carved wood of yesteryear in place of the molded plastic of today; we may yearn for the berries picked wild in the woods, but the rebellion is short-lived once we consider the facts.

There are already too many people on earth accustomed to too high a standard of living for us ever to return voluntarily to the simple and natural life of a bygone age. The good life which so many of us enjoy, at least in the United States and Western Europe, is geared inexorably to the assembly line and to the techniques of mass production, in agriculture as much as in industry.

In the days when food produce was grown naturally and goods were handcrafted of leather and wood, most of us would have been penniless serfs eking out a living in the service of the Lord of the Manor. Only a privileged few could enjoy handmade luxuries. The glories of Versailles, and the splendors of the Vatican, represent concentrated wealth; they were made possible by armies of peasant laborers, who themselves lived in hovels and mud huts, while creating architectural wonders enjoyed at the time by a minority and which have survived a thousand years. Plastic-molded mediocrity is the price we pay for shared wealth. We may sometimes reflect nostalgically that the price tag is high, but we would not change it for the world.

So the choice between a "natural" life and one dominated by science and technology is not the issue. There is no choice. A natural life is out of the question; science and technology are here to stay. The issue is to sort out the desirable from the undesirable.

THE INCREASED USE OF RADIATION

Our specific concern here is with the increased use of radiation in medicine, in industry, and in power production. Never has a new tool of mankind been introduced with such care. When coal was first burned in large quantities to produce power and to spark the Industrial Revolution, no committee of the National Academy of Sciences sat and deliberated on the possible health hazards caused by the smoke. When pesticides were first introduced, scarcely a thought was given to the possible deleterious effect to man or animal— there was too much jubilation that the scourge of insect pests had been controlled and food would be more plentiful.

Today we are more cautious. We have learned some lessons from the past, and if anything we are now downright skeptical. As a society we have learned that nothing is for free; we cannot expect to get anything without paying for it. There is always a price tag. In everyday life we are well used to the concept of paying for what we get, and of considering whether the cost is worth the benefit. This, essentially, is what we will have to do more and more in every facet of our daily lives. It is certainly what we must do in the case of radiation.

As we shall see in the chapters that follow, the exploitation of atomic radiations in medicine, industry, and power production can confer tremendous benefits. The exposure of the human race to too much radiation will undoubtedly produce misery and suffering. The question is—where do we draw the line? How do we balance the equation between benefit and risk?

2
The Invisible Rays

IONIZING RADIATIONS

So far in this book the word "radiation" has been used when what is specifically meant is *ionizing radiations*. A radiation is ionizing if it is able to disrupt the chemical bonds of molecules of which living things are made, and so cause changes which are biologically important. Visible light, radiowaves, as well as the radiant heat from the sun are also forms of radiation, but are not able to produce damage in this way by ionization, though, of course, they too can cause biological effects if larger amounts are involved.

Ionizing radiations come in several varieties.

First, there are rays, such as *x-rays* or *gamma rays*. These represent energy transmitted in a wave without the movement of any material, just as heat and light from the sun cross the vast emptiness of space to reach the earth. X- and gamma rays do not differ from one another in nature or in properties; the only difference between the two is where they originate. X-rays, in general, are made in an electrical device, such as may be seen in any dentist's office, while gamma rays are emitted by unstable or radioactive isotopes.

Other types of ionizing radiations are fast-moving particles of matter. Some carry a charge of electricity, some do not.

Neutrons are the only uncharged particles of any consequence, and are an important form of ionizing radiation because they are generally associated with atomic bombs and nuclear reactors. Neutrons are particles with a mass similar to that of the proton, but they carry no electrical charge. Because they are electrically neutral, they are very penetrating in material of all sorts, including living tissues. Neutrons constitute one of the fundamental particles of which the nucleus of all atoms is built. Neutrons are emitted as a by-product when heavy radioactive atoms such as uranium undergo fission, i.e., split up to form two smaller atoms. They can also be produced artificially by large accelerators in physics research laboratories.

Electrons are small negatively charged particles, which are found in all normal atoms. Electrons are often given off when radioactive materials break down and decay, in which case they are called *beta rays*. They can also be produced artificially in the laboratory and accelerated in electrical devices.

Protons are positively charged particles which are found in the nucleus of

The Electromagnetic Spectrum

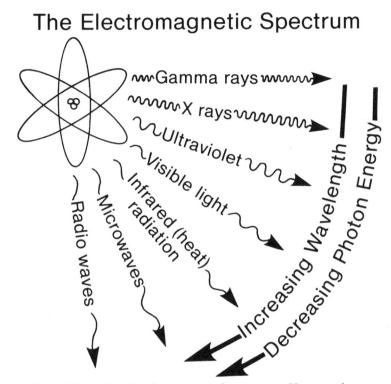

FIGURE 2.1. Illustrating the electro-magnetic spectrum. X-rays and gamma rays have the same nature as visible light, radiant heat and radiowaves; however, they have a shorter wavelength and consequently a larger photon energy—i.e., more energy per "packet."

every atom. They have a mass approximately equal to that of the neutron and are almost 2,000 times heavier than an electron. Protons are not usually given off by radioactive isotopes on earth, but they are found in great abundance in outer space and may constitute a hazard to astronauts.

Alpha particles are nuclei of helium atoms; that is, helium atoms with the planetary electrons stripped off. An alpha particle consists of two protons and two neutrons stuck together. They have a net positive charge and are relatively massive. These particles are commonly emitted when heavy radioactive isotopes, such as uranium or radium, decay and break down.

Heavy ions are the nuclei of any atoms which are stripped of their planetary electrons and which are moving at high speed. Ions of almost all of the known elements are present in space, and constitute one of the problems of space flight. These particles move with great speed and have enormous energy, and it is virtually impossible to design spacecraft to fully protect their occupants from all of the heavy ions.

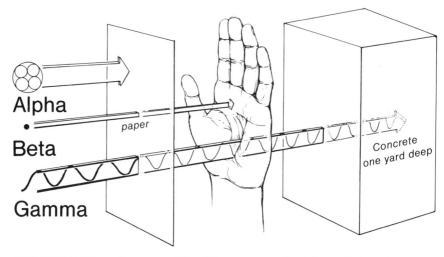

FIGURE 2.2. Illustrating some of the different types of ionizing radiations that were labelled α, β and γ before their true nature was known or understood. Gamma (γ) rays form part of the electromagnetic spectrum and are of the same nature as heat or light. They can be very penetrating and pass through thick barriers. Beta (β) rays are comprised of a stream of electrons—tiny negatively charged particles. Beta rays can pass through a hand, but unless they are of very high energy, they can be stopped by a modest barrier. Alpha (α) rays are relatively massive positively charged particles; they are in fact helium nuclei and each is made up of two neutrons and two protons in close association. In general alpha rays can be stopped by a thin barrier—even a sheet of cardboard.

ABSORPTION OF RADIATION

When such radiations pass through living things, they give up energy to the tissues and cells of which all biological materials are made. The energy is not spread out evenly, but is deposited or dumped very unevenly in discrete "packets." A lot of energy is given to some parts of some cells, and little, if any, to others.

This uneven pattern of energy deposition accounts for the special conse-quences of ionizing radiations. The total amount of energy involved may be small; however, some cells of the living material may be adversely affected because it is deposited so unevenly. The smallness of the amount of energy involved may be illustrated in a number of ways. For example, the dose of x-rays which would undoubtedly kill a man if given to the whole body may be compared with heat energy; it would be less than that absorbed by drinking a cup of warm coffee, or sunbathing for a few minutes on a hot day. Al-ternatively the energy inherent in a lethal dose of x-rays may be compared with mechanical energy or work; it would correspond to the work done in lifting a man about 16 inches from the ground.

Total Body Irradiation

Mass = 70 kgm
LD/50/60 = 4 Gy
Energy absorbed =

$$70 \times 4 = 280 \text{ Joules}$$
$$= \frac{280}{4.18} = 67 \text{ calories}$$

A

Drinking Hot Coffee

Excess temperature (°C) = 60° − 37° = 23°
Volume of coffee consumed to
equal the energy in the LD/50/60 $= \dfrac{67}{23}$

B

$$= 3 \text{ ml}$$
$$= 1 \text{ sip}$$

Mechanical Energy: Lifting a Person

Mass = 70 kgm
Height lifted to equal
the energy in the

$$\text{LD/50/60} = \frac{280}{70 \times 0.0981}$$
$$= 0.4\text{m (16 inches)}$$

C

FIGURE 2.3. Illustrating that the biological effect of radiation it *NOT* due to the *amount* of the energy absorbed, but due to the photon size or "packet" size of the energy. (a) The total amount of energy absorbed in a 70 Kilogram (154 lb) human exposed to a lethal dose of 4 Gray is only 67 calories. (b) This is equal to the energy absorbed when drinking 1 sip of hot coffee. (c) It also equals the potential energy imparted by lifting a person about 16 inches.

1 kg of sand
falling on a rabbit

1 kg of rock
falling on a rabbit

FIGURE 2.4. To illustrate the difference between ionizing and non-ionizing radiations, imagine throwing 1 kilogram of material at a rabbit. Thrown with the same speed, one kilogram of sand would comprise the same total amount of energy as a 1 kilogram rock. Yet, the sand would do little damage because the energy would be divided between thousands of tiny particles. By contrast, the rock could have a lethal effect if it hit a vital target, because the energy is concentrated in a large "packet." Similarly, ionizing radiations produce their biological effect, not because of the total amount of energy involved, but because the energy comes in large "packets," or photons, which are big enough to break vital chemical bonds (like the rock!). For non-ionizing radiations, such as visible light or radiant heat, the energy is divided into many small photons or packets (like grains of sand), which individually are not capable of breaking the chemical bonds necessary to produce serious biological effects.

Energy in the form of heat or mechanical energy is absorbed uniformly and evenly, and much greater quantities of energy in these forms are required to produce damage in living things.

The situation may be illustrated by an example which, although trivial, contains the essence of the difference between ionizing and nonionizing radiations. Imagine throwing a kilogram of material at a running rabbit. One may choose to throw sand, in which case millions of particles would make up one kilogram, or one may choose to throw a single rock weighing one kilogram. The same total amount of energy would be involved in throwing either projectile a given distance. In the case of the sand, the energy would be divided into such small individual packets that no damage would result to the rabbit from the impact of any of them. If a single rock were chosen instead, the chance of scoring a hit at all would be greatly reduced, but biological damage would undoubtedly follow in the event that a hit occurred.

Ionizing radiations constitute large discrete "packets" of energy. To absorb a dose of ionizing radiation is to be hit by a rock, not by many particles of sand!

MEASURING RADIATION

The amount or quantity of radiation, or as it is usually called, the dose, is measured in terms of the energy absorbed in the tissues. A number of units must be defined and explained in order to make the discussion of radiation effects comprehensible. An unfortunate additional confusion arises at the present time because the units that have been in use for many years are gradually being replaced by the internationally agreed "System Internationale" or SI units. The International Commission on Radiological Units and Measurements (ICRU) agreed that the new SI units for radiology would be introduced in 1980 and used side by side with the old units until 1984, when the old units would be phased out of usage. It must be admitted that the SI units, when introduced throughout science, are coherent and simple and will eventually be of benefit. In the transition period, however, the change causes as much confusion as did the decimalization of money in Britain in 1970 or the change to the metric system for weights and measures throughout the English-speaking world at the present time.

First, there is the unit of *absorbed dose*. The unit used for many years is the rad, which corresponds to an energy absorption of 100 ergs/gm. The new unit is the Gray (Gy), named after a famous British physicist and radiobiologist, which corresponds to 1 joule/kg. Therefore, 1 Gray = 100 rads. This is a physical unit, defined unequivocally, and does not allow room for debate or opinion. In some instances, the dose is very much less than one Gray or even one rad, and then the units milligray or millirad are used. There are 1,000 milligrays or one million micrograys in 1 Gray and 1,000 millirads or one million microrads in 1 rad. To give an idea of the way in which these units are used, two examples will be quoted from later chapters. A total body dose of 5 Gray (500 rad) to a human being would most probably be fatal; the average natural background radiation to which we are exposed is about 0.9 milligray (90 millirad) per year.

Equal doses of different types of radiation do not necessarily produce equal biological effects. 0.5 Gray (50 rads) of neutrons, for example, will be more effective than 0.5 Gray (50 rads) of x-rays. In general, x-rays, gamma rays and electrons are least effective for a given dose, while heavy ions are the most damaging. Neutrons fall somewhere in between.

For general discussions of radiation effects, a different quantity is used, namely *dose equivalent*, which is the absorbed dose (in Grays or rads) multiplied by a factor, known as the Quality Factor, which allows for the relative effectiveness of the particular type of radiation involved. If dose is measured in rads, dose equivalent will be in rems (rad equivalent man). If dose is in Grays, the dose equivalent will be in Sieverts, a new unit, named after a famous Swedish physicist who made substantial contributions to the methodology of measuring radiation. X-rays and gamma rays are regarded as the standard, and for these types of radiation rads and rems are interchangeable, as are Grays and Sieverts. A dose of 1 Gray (100 rads) of x-rays is, by definition, 1 Sievert (100 rems). Neutrons, on the other hand, are roughly 10 times more effective than x-rays and therefore are assigned a Quality Factor of 10.* Consequently, a dose of 1 Gray (100 rads) of neutrons represents 10 Sieverts (1000 rems). A dose equivalent in rems or Sieverts can only be approximate, since the factors needed to convert rads to rems or Grays to Sieverts are not known with great accuracy for all types of radiation, and in any case may differ for various biological systems. In some instances, the dose equivalent may be much less than one Sievert or one rem, in which case smaller sub-units are used. There are 1,000 millisieverts and one million microsieverts in one sievert. There are 1,000 millirems and one million microrems in one rem.

Next, there is the *collective dose equivalent*. This is obtained by multiplying the average dose equivalent by the number of individuals exposed. Thus, for example, in the "incident" at Three Mile Island, an estimate of the collective dose equivalent received by the 2 million people who lived within 50 miles turned out to be 32 person-sieverts (3,200 man rems). Some individuals received up to 1 millisievert (100 millirems), others less than .01 millisievert (1 millirem).

Finally, there is the *committed dose equivalent*. This is the estimate of the dose that will be received in the future by the human population from the release of a certain amount of radioactivity. The estimate must take into account the physical decay of the radioactivity, the number of people that may be affected in the future, as well as the rate at which the radioactivity

---

*There is discussion at present of increasing this factor to 50.

may be taken in and excreted by human beings. The units will be the same as collective dose equivalent, i.e., person sieverts (man rems).

Activity of Radioactive Materials

The quantity or amount of radioactive material is specified in terms of its *"activity"*; it would make no sense to speak in terms of its weight. The old unit used for many years is the Curie, named in honor of Pierre and Marie Curie, who first isolated radium. Historically, the unit is based on radium for which 1 gram has an activity of 1 Curie. When applied to all radioactive materials, the Curie is defined as that amount of material in which 37 thousand million atoms disintegrate per second. Since the Curie is a relatively large unit, the millicurie (one thousandth of a Curie), the microcurie (one millionth of a Curie), and the pico Curie (one millionth of a millionth of a Curie) are more frequently used.

The new unit for activity in the SI system is the Becquerel, named after the discoverer of radioactivity, and defined to be 1 disintegration per second. The Becquerel is small compared to the Curie; one Becquerel equals 27 pico-curies. The change of units caused some anger among those female scientists who support the women's liberation movement, since it removes from common usage the one unit named after a woman!

It is important to recognize that a statement of the activity of a radioactive isotope (in Becquerels or Curies) does not per se give any indication of the radiation dose (in Grays or rads), much less of the dose equivalent (in Sieverts or rems). The activity only specifies the number of atoms disintegrating per second, and gives no indication of the type of radiation emitted nor its energy. Consequently, the same activity does not imply the same degree of hazard for different radioactive materials. For example, 37 megabecquerels (one milli-curie) of polonium could be extremely hazardous, whereas the same activity of tritiated water is quite innocuous.

Decay of Radioactive Materials

The time taken for the activity of a radioactive material to lose half its value by decay is called the *half-life*. Each radionuclide has a characteristic and unalterable half-life, which can range from seconds to millions of years. Uranium-238, for example, decays with a half-life of 4,470 million years, while for Iodine-131 the half-life is only 8 days. In successive half-lives, the activity is reduced to $\frac{1}{2}$, $\frac{1}{4}$, $\frac{1}{8}$, $\frac{1}{16}$ and so on of its initial value, so that it is possible to calculate and predict the amount that will remain at some future time.

The New Units

The confusion caused by the change of units has inspired a number of would-be poets, the efforts of one of which is reproduced below.

The New Units

The Rad will soon have had its day,
and then for dose we use the GRAY.
But Grays are bigger as we've said
So talk of centi-grays instead,
That brings us right back to the start,
You can't tell these and rads apart.

The REM for dose equivalent
Has been replaced, so don't relent
The SIEVERT is the newer unit
A hundred times as large, so prune it,
Down to centi-sievert which
Equals a rem—that's quite a switch.

The dropping of the former CURIE
Puts women's libbers in a fury
Soon we must use the Becquerel
Which few of us can even spell
But one Becquerel plus women's fury
Makes twenty-seven pico-curie.

Based on a poem by
Angela Newing
Editor, *The Hospital Physicist's Association Bulletin*

Table 2.1. Relation of Dose Units and Quantities

ABSORBED DOSE

Energy absorbed from the radiation per unit mass of tissue (Gray, rad)

↓

DOSE EQUIVALENT

Absorbed dose multiplied by the quality factor in order to allow for the different effectiveness of different types of radiation (sievert, rem)

↓

COLLECTIVE DOSE EQUIVALENT

Dose equivalent to a group or population from a radiation exposure, calculated by multiplying the average dose equivalent by the number of people exposed (person-sievert, man-rem)

Postscript on dose units

The units of absorbed dose (rad or Gray) represent units of physical quantities which are rigorously defined and leave no margin for debate or opinion. On the other hand, dose equivalent (rem or Sievert) includes a value for the Quality Factor which is decided upon by an International Committee. Quality Factor values are best estimates based on experimental data, but their exact values are a subject of some uncertainty and debate. The collective dose not only requires acceptance of Quality Factor, but involves the assumption that the biological effect is a linear function of dose; i.e., that the effectiveness per unit dose is independent of the dose level. For example, a million people exposed to one microsievert or 1,000ʹ people exposed to one millisievert would both constitute a collective dose of one person Sievert. It is by no means certain or agreed upon that the same biological consequence would result from the two disparate cases. It is a useful concept for assessing the impact of a radiation exposure, particularly in the context of an accident, but it must clearly be taken with a generous pinch of salt. It is a quantity much loved by the paper-pushers in the government bureaucracies, since it makes the estimate of the impact of a radiation exposure on the public simple and unequivocal, even though it may have little or no meaning. Collective dose has little meaning at all without introducing the concept of *"de minimus"* dose. The idea is to define some very low dose below which it is assumed that there will be no effect. Unless this is done, the concept of collective dose becomes unworkable. For example, suppose a release of radioactivity occurs in the U.S. and results in a small exposure to a few individuals. The airborne radioactivity may blow around the world, being diluted progressively until it reaches China. By this time, the exposure of any individual may be only a few millionths of a rem and totally unmeasurable, but when multiplied by the vast number of people in that country it would dominate the dose commitment. The same principle of course applies to other potentially harmful pollutants—for example, a puff of smoke released from a chimney in Chicago may eventually reach Australia, and while it may be possible to calculate the concentration there, it would have little meaning and the effects would be so small that they could be neglected. The term *"de minimus"* comes from the legal—*De Minimus non curat lex*, which roughly translates to: "The law does not concern itself with trifles."

Dr. Merril Eisenbud in an NCRP publication quotes this poem of dubious origin:

> "There was a young lawyer named Rex
> Who was very deficient in sex
> When charged with exposure
> He said with composure
> De Minimus non curat lex."

EXTERNAL VERSUS INTERNAL HAZARD

It is fundamental to the understanding of the radiation hazard to realize that there are two distinct ways in which ionizing radiation can reach and affect the tissues of the body. First, there is external radiation which comes from a source outside the body. The radiation, like x-rays, or gamma rays in this case, must be relatively energetic in order to penetrate through the body, or like some energetic beta rays be able to penetrate the superficial layers of the skin. Second, there is internal radiation resulting from radioactive materials which have gained access to the interior of the body. Alpha, beta and gamma-rays can be a problem in this respect, but the most serious problem comes from the deposition of isotopes that emit short-range densely ionizing alpha particles. It stands to reason that the precautions that must be taken against the external radiation hazards are quite different from those associated with internal radiation hazards.

EXTERNAL RADIATION

External radiation hazards can be categorized as either highly penetrating radiation consisting of x-rays, γ-rays, or neutrons and less penetrating radiation such as energetic beta rays or electrons. Highly penetrating external radiation, such as x- or γ-rays and neutrons, can reach, and therefore affect, all of the tissues and organs of the body. The means of protection from external x or gamma and neutron radiations therefore depends on a combination of three factors: time, distance, and shielding. That is to say, protection is achieved by controlling the length of time of exposure, the distance between the individual exposed and the source of the radiation, and interposing an absorbing material between the individual and the source of the radiation.

The effect of *time* on radiation exposure is relatively easy to understand. If an individual is situated in an area where the radiation level from penetrating external radiation is 10 millirem/hr, then in 1 hour he would receive 10 millirem of exposure, if he stayed 2 hours he would get 20 millirems, if he stayed 4 hours 40 millirems, if he stayed 8 hours 80 millirems, and so on. In other words there is clearly a linear relationship between time and the dose absorbed. Time is thus used as a factor in keeping the overall exposure down to the absolute minimum. For instance, if work must be performed in a high-radiation area, then the work to be done must be carefully planned in advance so that the time in the radiation area necessary to accomplish the task is minimized. By contrast, the effect of distance on radiation exposure is quite startling, since the dose rate falls off according to the inverse square law. That is to say, the intensity of radiation falls off by the square of the distance from the source. For example, if there is a point source of radiation giving 100 rems/hr of penetrating external radiation at 1 meter, then at twice the distance, or 2 meters, the intensity of radiation would be reduced by a factor of 4 to only 25 rems/hr. When the distance is tripled, the dose rate would be reduced to one-ninth of the dose rate, and so on. There is, then, a dramatic fall-off in the rate of radiation exposure as distances are increased from the source of radiation resulting in a very effective means of protection from penetrating x- or γ-rays and neutrons.

The discussion of shielding is a little more complicated. The effectiveness of a material as a shield or screen against penetrating x- or γ-rays depends upon the density of the material and upon its electron density as well. Materials such as lead or uranium are very much more effective as shields than materials such as aluminum, water, or paper. On the other hand, protection against penetrating neutrons is most effectively achieved by the use of materials that contain a lot of protons such as water or paraffin wax.

Because of the way in which radiation is absorbed, the thickness of the shield does not have to increase in direct proportion to the amount of radiation being shielded. For instance, a container for a 1,000 Curie radiation source does not have to be 1,000 times as heavy as a container for a 1 curie source. As a matter of fact, the container need only be about 13 times as heavy for the larger source as for the smaller source. If one half-value layer (HVL) of shielding is added, the intensity of the radiation emerging on the other side of the shielding is reduced by a factor of 2. Two HVL layers will reduce it by a factor of 4, 3 HVL layers will reduce it by a factor of 8, 4 HVL layers will reduce it by a factor of 16, and so on.

INTERNAL RADIATION PROBLEMS

The problems involved with internal radiation exposure are much more complicated than those involving external radiation exposure. There are essentially four possible ways in which radioactive materials may enter the body: (1) by inhalation; (2) by ingestion; (3) through breaks or cuts in the skin; (4) by absorption through the intact skin.

In the case of radioactive material that is inhaled, the very small particles will be exhaled, while larger particles are trapped in the hairs and mucous passages and eventually blown out. Certain sized particles will be trapped in passages of the lung. If the particle is insoluble, it remains in the lung and the radiation dose is delivered to the lung tissue. In particular, if the isotope emits short-range alpha particles, enormously high local radiation doses can then be delivered to certain tissues in the lung. On the other hand, if the particle is soluble, it will get into the blood stream and be transported to various tissues and organs of the body.

Materials taken in through the skin go directly into the blood stream, and the fate of the radioactive isotope depends upon its chemical properties. Certain materials are preferentially absorbed and taken up in specific organs leading to high local radiation doses; however, if they are not incorporated into the tissues and organs of the body they will eventually be disposed of through the kidneys and excreted in the urine. For instance, the bones incorporate calcium, and since radium is in the same group in the periodic table as calcium, when radium is ingested it is preferentially deposited in the growing tips of the bones and can lead to very high local doses because of the emission of alpha, beta, and gamma-rays. Another example is radioactive iodine, which is preferentially taken up in the thyroid and can result in large local doses because of the emission of beta and gamma rays. Some chemical substances, such as sodium and potassium, are widely used throughout the body and therefore if a radioactive form of one of these elements is introduced into the body it will also be dispersed throughout the entire body. Other elements tend to concentrate in specific organs, as iodine in the thyroid gland. The point to remember is that body organs react to a substance on the basis of its chemical nature alone, without regard to whether or not it is radioactive.

Once deposited, the important factors are the energy and range of the radiation, shape and weight of the organ, and the physical and biological half-life of the material. The half-life is the amount of time that it takes half of the material to lose its radioactivity. If a material has a radioactive half-life measured in fractions of a minute, for instance, it will be dissipated very rapidly. On the other hand, if the material has an extremely long half-life, possibly measured in thousands of years, then the rate at which it decays is very slow. The biological half-life of the material is that period of time which it takes for half of the material to be excreted from the body. Some materials are excreted quite rapidly from the body and therefore will not be in it long enough to do much harm, while others remain over a considerable period of time. The combination of the physical half-life of the radioactive material and the biological half-life governed by the processes of excretion in the body lead to the *effective half-life* of the material in the body, which is one of the most important factors in determining the resultant amount of radiation exposure to the tissues of the body. The organ that is most affected by the radioactive material is the *critical organ*. Standards have been developed for the amount of each radioactive material to be permitted in the various critical organs, and based on these figures are the standards allowed for the amount of radioactive material permitted to be airborne or dissolved in drinking water, etc. Protection from internal radiation hazards is based principally on maintaining good techniques over an extended period of time.

RADIATION PROTECTION GUIDES

It is common practice with toxic substances to establish limits for exposure to the public that are small fractions of those which apply to people who are exposed in the course of their daily work. The justification of this is that greater risks are acceptable for a small group of people in a specific industry or profession, than would be tolerated if applied to the whole population comprised as it is, of many millions of people. These limits are known as "standards" or "protection guides."

Doctors and scientists who work with x-rays in hospitals, and workers in many other areas of the nuclear industry are occupationally exposed to radiation. Some radiation exposure in the course of performing their job is generally unavoidable. Under these conditions the maximum permissible dose equivalent is 30 millisieverts (3 rem) in any calendar quarter, with the added requirement that the cumulative dose equivalent must not average more than 50 millisieverts (5 rem) for each year that the person's age exceeds 18. The maximum dose equivalent in rem cannot exceed 5 (N-18) where N is the age in years. An 18-year-old person would be restricted to an exposure of 50 millisieverts (5 rem) per year, but for an older person who starts working with radiation later in life, the annual limit would be 120 millisieverts (12 rem) per year, provided not more than 30 millisieverts (3 rem) were received in any one quarter. If the quarterly limit is exceeded, it must be reported, by law, to the U.S. Nuclear Regulatory Commission or a local regulatory agency. Such an event could result in a significant fine. In the case of the medical use of radiation, the group of workers who receive the largest doses are those engaged in fluoroscopy (see chapter 6). These individuals may receive up to 5 millisieverts (500 millirem) per year. Ninety-five percent of the radiation workers in hospitals receive less than 3 millisieverts (300 millirems) per year and many receive amounts too small to be measured. In the nuclear power industry, the average annual dose received is about 8 millisieverts (800 millirem), but there is a small group of skilled individuals responsible for the maintenance and repair of nuclear reactor components who receive close to the maximum permissible levels, 50 to 120 millisieverts (5 to 12 rem) per year. It should be emphasized that the dose limits apply to occupational exposure. When the same individual goes to the doctor or dentist and is exposed to x-rays as a patient for the benefit of his own health, these doses are exempt from record keeping of occupational exposure. The occupationally exposed worker *removes* his monitoring badge whenever there is a potential for medical exposure.

BURN-UP

This term, often used by the media and popular press, conjures up visions of individuals being scorched and burned by excessive doses of radiation. In fact it refers to the practice whereby radiation workers receive the full quarterly allowance of 30 millisieverts (3 rem) in a few minutes. If repairs are required, for example, on the valves or pumps in the primary cooling circuit of a nuclear reactor, it is inevitable that the individuals involved will receive relatively high levels of radiation. The health physicist measures or calculates the dose levels involved and successive technicians work in turn for brief periods of time so that no single individual receives more than the allowable dose. In this way, the work is done without exceeding the regulatory limits! Private enterprise in the United States has led to the setting up of "Body Shops," pools of experienced radiation technologists who receive little radiation in the course of their regular jobs during the year, therefore they may choose to work for a few weeks during their vacations at nuclear power plants during outages. The financial reimbursement for their services provides the incentive for registering with these "Body Shops." It would be economically unreasonable to require the nuclear power industry to operate without procedures such as this; it differs little from paying high wages to steeplejacks for the risky job of working on top of tall buildings. It must be admitted that the practice of "Burn-Up" carries with it the risk of misuse and exploitation. For example, it is alleged that when nuclear powered submarines or aircraft carriers are refueled by navy technicians, no expense is spared to build shielding walls and minimize the irradiation of personnel. By contrast, the same procedure in private enterprise shipyards is done as quickly and as inexpensively as possible, with less regard for protection, by using teams of workers who receive their quarterly allowance of 30 millisieverts (3 rem) and are then replaced by others; expense is spared at the cost of irradiating more people with larger doses.

For the public, *individuals* can be exposed to 5 millisieverts (0.5 rem) per year; one-third of this value, 1.7 millisieverts (0.17 rem) per year applies to large *groups* of people. This figure of 1.7 millisieverts (0.17 rem) per year for the general public has also been derived in a totally different way, which in many respects is more relevant. In the mid-1950s the National Academy of Sciences established a series of committees, the so-called BEAR committees, to investigate the biological effects of atomic radiations. In 1956 the committee on genetic effects issued its report. This committee recommended that, on the basis of potential genetic effects, the total population should receive no more than 0.1 Sievert (10 rems) over a 30-year period, which is taken as the mean reproductive age of the human being. This dose was intended to apply to exposures from all man-made resources, including radiation used in medicine. Half of this value, or .05 sieverts (5 rems) over 30 years, was later allocated to all sources *other* than medical. This leads to 1.7 millisieverts (0.17 rem) average dose per year from man-made sources of radiation, other than medical.

In the ensuing years, emphasis has shifted from the genetic effects of low doses of radiation to the production of cancer and leukemia, which is now considered to be the principal hazard of low doses. In spite of this fundamental shift in the biological basis, the radiation protection guidelines and maximum permissible doses have not changed, and the surveillance of the large number of workers occupationally exposed indicates that there is no need for it to change.

This guideline obviously represents a value judgment. It is equal, approximately, to the amount of natural background radiation that human beings receive, a fact which played a significant role in the derivation of this figure. The reason for this is that background radiation represents an exposure which human beings have experienced over eons. Living things evolved from the most primitive stages while being exposed continuously to background radiation levels that were probably much higher than those at the present time. In addition, the amount of background radiation varies considerably over the face of the earth. In large areas of France, the background radiation averages about twice the levels in the United States. In some parts of India, very large populations are exposed to levels of background radiation 10 times that which is experienced in the United States, with no noticeable detrimental effects. Consequently, committees whose duty it is to suggest radiation standards feel confident about protection levels that are *similar* to the doses administered by background radiation, but feel rapidly less secure as exposure levels *exceed* these figures.

AS LOW AS REASONABLY ACHIEVABLE (ALARA)

The committee which set the number of 0.17 rem (1.7 millisieverts) per year as an upper limit of exposure of the general population made it clear that they were not necessarily implying that there would be no harm to the population at these dose levels, or that such dose levels are "safe." They did say, however, that they felt confident that at levels near background exposure the effect on the population, if any, would be quite small and that such doses would not represent a threat to the continued existence and propagation of the human population.

The setting of this standard was not the only important recommendation of the BEAR Committee. Although they recognized that there may well be *no* harmful effect at very low doses, they accepted the thesis that until there was concrete evidence to the contrary, it was prudent to assume that any amount of radiation exposure may carry some probability of harm to the population, no matter how small that probability may be. The recommendation that the committee would have preferred to make is zero exposure. However, they realized that zero exposure is not only impossible but also impractical. Exposure from natural sources is unavoidable, and some additional exposure from man-made devices is unavoidable if man is to realize the enormous benefits derived from the uses of radiations and radioactive materials.

Radiation is potentially harmful, and exposure to it should be continually monitored and controlled. No unnecessary exposure should be allowed; the best radiation protection guide is to keep radiation exposures "as low as reasonably achievable." Equipment and facilities should be designed so that exposure of personnel and the public should be kept down to a minimum and not up to a standard. No exposure at all should be permitted without considering the benefits that may be derived from that exposure, and without considering the relative risks of alternative approaches.

Of course, the ultimate problem is what is "reasonable." How much expense is justified to reduce the exposure of personnel by a given amount? In the nuclear power industry in the United States, ALARA has a cash value which is currently about $1,000 per rem (10 millisieverts). If the exposure of one person to 10 millisieverts (1 rem) can be avoided by the expenditure of less than $1,000, then it is reasonable. If the cost is more, it is considered to be unreasonable and the exposure is allowed.

3
Of Cells, Mice and Men

THE BASIC BIOLOGY

All living things are composed of cells, which are the basic building bricks of life. Every part of the body, from the brain to the blood, and from the bones to the glands, is made up of colonies of cells. In an adult organism, each and every cell has a set of chromosomes that contains, in code form, the complete genetic information to specify the size and shape and nature of that organism. Every one of the billions of cells in an elephant contains the information to make an elephant; every cell in the mouse contains the information to make a mouse.

The number of chromosomes per cell varies with the species; in the human, the normal cell contains 46 chromosomes, and most other higher mammals contain a similar number. The marsupials, including the kangaroo, the wallaby, and the opossum, have a small number of chromosomes, sometimes as low as 11. In plants, the number of chromosomes and their size is very variable, but in general they have a smaller number of chromosomes, which are large and simply formed.

In the case of higher animals, the first cell of a new creature comes into being at the moment of conception, when a sperm and an egg cell unite. These germ cells have a number of special features, most notable of which is that they contain only *half* the number of chromosomes found in all the other cells of the body. In this way the genetic material of the offspring is made up of half from each parent. For example, the first cell of a new human being contains 23 chromosomes donated by the mother and an equal number from the father.

This single cell, produced at conception, then replicates or divides to form two cells, each of which divides again thus forming four cells, and so on. At this early stage, all of the cells are capable of division, and so the new cell population increases at a very rapid rate. At a later stage, some of the cells "differentiate" and become specialized, capable of performing a specific function, but unable to divide again. Some become brain cells, some liver cells, and so on. In the course of time, the mature adult organism is formed, composed of populations of dividing and differentiated cells, every one of which bears a legacy of the genetic characteristics of both parents.

BIOLOGICAL EFFECTS OF RADIATION

The effects of radiation on living things, particularly human beings, may be expressed in quite different ways depending principally on the size of the dose. They may be listed as follows:

1. Changes in somatic cells that cause cancer.
2. Genetic mutations that affect future generations.
3. Effects on the embryo and fetus due to irradiation during pregnancy.
4. Immediate radiation death.

The first two, changes of body cells that cause cancer or mutations of the sex cells which affect future generations, are the biological consequences of most concern when *large* numbers of people are exposed to *small* doses of radiation either for medical purposes or as a result of nuclear power stations. Exposure of the developing embryo or fetus is a special case that deserves a detailed discussion because every effort should be made to avoid it. The fourth effect of radiation, namely immediate death, requires enormous doses. It only occurs in a catastrophic situation, such as the explosion of an atom bomb or an accident in a nuclear reactor.

These various consequences of radiation will now be discussed in turn.

1. Changes in Somatic Cells that Produce Cancer

Radiation can produce cancer in animals and in man. This is a fact beyond dispute. A cancer is produced when a somatic cell goes berserk, ceases to obey the controls of the body so that it divides again and again with no regard for the well-being of the creature as a whole, and forms a single large mass or series of masses. The initial event which causes the somatic cell to behave in this way is probably a change in its genetic apparatus. This is called a mutation. A mutation or change in a germ cell will affect future generations; a mutation in a somatic cell has consequences only for the individual concerned. The cancer cell reproduces itself by dividing into two cells, which in turn redivide and so on. All of the descendants of the original cancer cell are themselves cancer cells. In this way, the tumor grows.

Radiation-Induced Cancer in Humans

There are many instances where cancer has been produced in humans as a result of radiation. This experience will be summarized briefly.

The early radiologists and scientists who worked with x-rays at the turn of the century had no means of knowing that their newly discovered rays were dangerous, and took few precautions to safeguard themselves. Many died from leukemia and cancers of the skin and bones. Marie Curie, for example, who first isolated radium and polonium, died of leukemia, as did her daugh-

FIGURE 3.1. Marie Curie (seated) at work with her daughter Irene. Both were eventually to die of leukemia induced by the phenomenon for which Marie devised the name radioactivity. (The photograph was made available by the Austrian Radium Institute and published by the International Atomic Energy Agency Bulletin.)

ter Irene who helped her mother in her research on radioactivity. The young factory girls who, in the 1920s, painted watches with luminous dials, often licked their paintbrushes to give them a sharp point. The luminous paint contained radium, which was ingested and deposited in the growing tips of their bones; the radiation from these active deposits produced tumors which became evident many years later. Almost every girl traced who did not die of anemia or its effects, died later of bone cancer.

The pitchblende miners in Austria and uranium miners in Colorado used to inhale continuously a radioactive gas, called radon, in the course of their

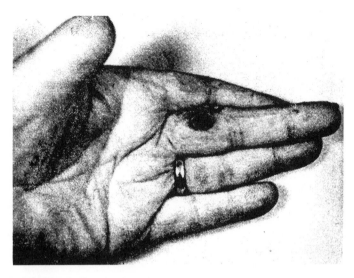

FIGURE 3.2. Hand of a dentist who for 35 years had held x-ray films in place in patients' mouths. The thumb has been partially amputated; damaged skin on the fingers has been replaced by grafts; the lesion on the finger is a skin cancer subsequently removed. (Courtesy of Dr. Victor Bond, Brookhaven National Lab.)

work. Radioactive deposits built up in the lungs and as a consequence the death rate of these miners from lung cancer is much higher than in the rest of the population.

Leukemia is a disease characterized by a great excess of abnormal white cells in the blood; it is often called "cancer of the blood." The circulating white corpuscles do not themselves divide, but are produced from actively dividing stem cells in the bone marrow and lymph glands. A change in one or more of these stem cells floods the body with abnormal white cells, and this is leukemia. A link between radiation and leukemia is well proven. There are several pieces of incontrovertible evidence.

In Great Britain during the 1930s and early 1940s, more than 14,000 patients were treated with large doses of x-rays for a painful disease of the spine known as ankylosing spondylitis. The treatments were very successful, pain was relieved, and the patients lived for many years. When their records are examined in retrospect, it is evident that leukemia was the cause of death in more of these patients than in a comparable number of people from the general population. The numbers were not large; in nearly 14,000 patients treated, about 70 developed leukemia, whereas only about two cases would be expected in the same number of people not irradiated. Even with this knowledge, most patients would still opt for the radiation treatment, enjoy certain and lasting relief of their pain, and take the risk of being the one in every 200 doomed to die later of leukemia. The odds are not outrageous for someone who is ill and in pain.

Some of the most reliable data on radiation-induced cancer in the human comes as a concomitant to the medical uses of radiation. There are three instances. First, it was common practice in the 1930s and 1940s to use x-rays to treat children supposed to have an enlarged thymus, a gland located in the chest. Incidental to this treatment, a substantial dose of radiation is given to the thyroid, and when these children are examined many years later it is found that a small proportion of them have either benign or malignant tumors of the thyroid. Second, x-rays were widely used in the 1940s and 1950s for the treatment of ringworm of the scalp (*tinea capitis*). In resistant cases of this disease, a dose of about 4 Gray of x-rays (400 rad) was given to the head of each affected child, which caused the hair to fall out, so that the ringworm in the hair follicles could be treated and eradicated with the drugs available at that time. The radiation dose used produced only temporary epilation and the hair grew back again within a few months. Doctors in Israel treated thousands of children in the years after the establishment of the State of Israel when ringworm of the scalp reached epidemic proportion amongst immigrants from North Africa. They noticed a small proportion of children treated in this way subsequently developed thyroid cancer. Following this initial observation, similar reports have been published from various parts of the world. This is another instance when exposure of the thyroid gland was inherent to the treatment of a particular disease. In these cases, it is estimated that the thyroid gland would receive between 0.03 and 0.3 Gray (3 to 30 rad), which represents probably the lowest dose of radiation at which cancer has been observed in the human. From these two examples, in which thyroid cancer is associated with a medical use of radiation for the treatment of some of the sites in the body, come our best estimates of the risk of radiation-induced thyroid cancer. It turns out, as will be discussed later, that the thyroid is one of the most sensitive tissues in the body for the induction of cancer by radiation. It is interesting to note that a large series of children in New York, similarly treated for ringworm, developed benign thyroid nodules, but *no* thyroid cancer. On the other hand, cancer was observed in areas of the skin regularly exposed to sunlight, such as the face. This was noticed in the white, but not in the black children treated. Skin cancer was not observed in the Israeli children treated, presumably because most came from North Africa and were dark skinned.

The third instance concerns the link between fluoroscopy and cancer. During a fluoroscopic examination, the patient is irradiated with x-rays for several minutes while the doctor looks at a sensitive screen on which the x-ray image of the chest is projected in order to look for lesions and deformities. Women with tuberculosis who received fluoroscopic x-ray examinations many times during the management of their disease showed an elevated incidence of breast cancer as a result. This experience was first reported from a TB sanitarium in Nova Scotia, and has been subsequently confirmed by similar studies in the New England states. This is one of the few examples available where a definite increase in cancer incidence has been observed following the use of diagnostic x-rays, though it must be admitted that these patients received substantial doses of radiation resulting from, in some cases, 100 to 300 fluoroscopic examinations.

The survivors of the A-bomb attacks on Hiroshima and Nagasaki form the largest group of human beings exposed to whole-body radiation. These people have been very carefully followed and studied in the years since the war, although it is difficult to assess who was and was not irradiated, and even more difficult to estimate the doses they received. In addition, many of the radiation cases are complicated by the more conventional aspects of the explosion.

The radiation aspects of the bomb have aroused a violent emotional reaction that is understandable because radiation effects are new, and fear is always generated by the unknown. However, to put things into perspective, it should be noted that most of those who died at Hiroshima were victims of the blast, fire, and debris from falling buildings, and died in ways no different from the countless air raid victims in Cologne or Essen, in London or Coventry. Indeed, the loss of life from these unique atom bomb attacks were considerably less than from the fire raids on Tokyo a few months earlier, or on Dresden in the previous year.

Radiation Induced Leukemia

Nevertheless, there is no question that some survivors of the bomb died of leukemia. The number of excess deaths rose to a peak by the early 1950s, that is 5 to 7 years after the explosion, but by 20 years no additional new cases were being reported which could be attributed to radiation. Of all the malignancies caused by radiation, leukemia is the one we know most about, because the time between cause and effect is comparatively short. What are the numbers involved? Of the 113,006 people who survived, and were irradiated, it is estimated that about 100 have died of leukemia resulting from the radiation. This figure cannot be precise, because in any group of over 100,000 people some would get the disease naturally. The incidence of leukemia in survivors was related to their distance from the explosion, and thus to the dose of radiation received. The highest incidence was close to the explosion, and the lowest at a greater distance. This is clear evidence that the production of the disease was due to radiation, since it is dose dependent.

While it is leukemia that is associated with the bomb in the minds of most people, it is becoming apparent as the years pass that it is not the most important form of cancer produced. As the Japanese survivors are studied

FIGURE 3.3. (*Top*) "Little Boy," the A-bomb dropped on Hiroshima on 6th August, 1945, was a gun-assembled uranium design (length 120 inches, diameter 28 inches). There was no test firing of a similar device. (*Bottom*) "Fat Man," the A-bomb dropped on Nagasaki on 9th August, 1945 was an implosion-assembled plutonium weapon (length 128 inches, diameter 60.5 inches). Several similar devices were tested before and after.

for a longer period of time, lung cancer, breast cancer, and particularly thyroid cancer, all appear more frequently than in the Japanese population as a whole. These types of cancer take much longer to develop; some are only

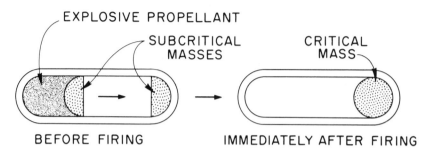

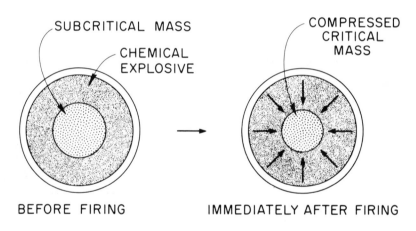

FIGURE 3.4. An atom bomb consists of two masses of fissionable material, such as uranium-235 or plutonium, each of which is too small to go "critical" and sustain a chain reaction i.e. subcritical masses. When they are brought together, they constitute a critical mass and the bomb "explodes." There are various ways to achieve this. The figure is a schematic illustrating the gun-assembly nuclear device (Little Boy) used at Hiroshima and the implosion-assembly device (Fat Man) used at Nagasaki.

appearing now, 40 years after the bomb. At the present time, there are about three cases of solid tumors for each case of leukemia produced by radiation. The number is still increasing, and by the time the A-bomb survivors have lived out their lifespan this figure may be as high as five.

One last point must be made regarding the oft-studied group of radiation martyrs, the Japanese survivors. These unfortunate folk received a sudden burst of radiation, which was a mixture of gamma rays and neutrons. The dose received by any individual can only be estimated very roughly from their retrospective account of where they were at the time of the explosion, and whether or not they were shielded by a building of wood or of brick. The data are subject to all sorts of uncertainties, and this should always be kept in mind.

In Utero Irradiation with Diagnostic X-Rays

Dr. Alice Stewart and her colleagues did a careful retrospective study of the records of 7649 children in Britain who died before the age of 10 of leukemia or some other childhood cancer. They found that, of those who died, 1,141 had been x-rayed while in utero, i.e., before they were born and while their mothers were pregnant. Of an equal number of children who did not contract cancer, only 774 received x-rays. The x-ray examination involved one to five films with a dose to the developing embryo or fetus of 2 to 4.5 milligray (0.2 to 0.45 rad) per film. Dr. Stewart concluded from these figures that x-rays produced the excess cancer, and for this to be the explanation it would require the unborn child to be extremely sensitive to radiation-induced carcinogenesis—much more so than the adult. This is a possible explanation and has received a great deal of attention in the press, but it is not an inevitable conclusion from the study. All the study demonstrates unequivocally is that those who died of cancer had a bigger chance of having had an x-ray, but it doesn't identify radiation as the cause of the cancers. There is the alternative explanation, that those who received x-rays in utero represent a selected group predisposed to cancer. Their mothers experienced problems during pregnancy, sought medical help, and received x-rays as part of the diagnostic procedure. This alternative explanation receives support from two sources. First, a later study showed that the brothers and sisters of the children who died of cancer had *twice* the cancer incidence of normal children. This serves as strong evidence that there is a familial predisposition to cancer and that Alice Stewart's study was dealing with a selected group. The other piece of evidence comes from the Japanese survivors. Over 2,000 children were born of mothers who were pregnant at the time of the atomic bombing; they received in utero an average dose of about 0.14 Sievert (14 rem)—a quite significant dose and much more than the children in Alice Stewart's study. If Alice Stewart were correct in assuming a high sensitivity for the unborn child to the production of cancer by radiation, 12 cases would have been expected in the 2,000 children of Hiroshima and Nagasaki, when in fact only one was observed—almost exactly what would be expected naturally from the average figures for the whole of Japan without any effects of radiation.

It appears then that the conclusions of Alice Stewart's famous study do not stand up to more careful scrutiny. No one denies the observation, which has been repeated in similar studies in the United States, that children x-rayed before birth have an elevated incidence of childhood cancer. But the excess cancer is due to the selecting of a group predisposed to cancer, not to the extreme sensitivity of the unborn child to carcinogenesis by low doses of radiation produced by several diagnostic x-ray films. There appears to be no evidence at the present time that the unborn child is *more* sensitive to carcinogenesis than the adult.

Lessons from the Human Experience of Carcinogenesis

A number of lessons can be learned from this brief summary of the human experience of radiation induced carcinogenesis. In the first place, most of the early examples of carcinogenesis resulting from radiation involved instances that are of historical interest but because they are anecdotal and because the radiation exposure took place under conditions that are not known in detail, it is not possible to infer any quantitative relationship between the dose of radiation absorbed and the number of cancers produced. It is also of limited usefulness because the situations involved no longer represent a public health hazard. For example, the fact that miners at the turn of the century developed lung cancer by breathing radon in mines that were not ventilated, is anecdotal and historical, and will never happen again because modern uranium mines are adequately ventilated to prevent the accumulation of a concentration of radon. More than that, it is impossible to estimate the concentration of radon at the bottom of a mine in Saxony at the turn of the century, so that we do not know the doses of radiation involved in the production of the lung cancers in these miners. As a result, the information is of limited usefulness. It is only the later examples, starting perhaps with the ankylosing spondylitic in the 1930s, where the radiation exposures were received under conditions sufficiently controlled, that a quantitative estimate can be made of the number of cancers produced as a function of the radiation dose absorbed.

An important lesson that we can learn from this summary is that, while the cancers produced in the Japanese from the atomic bombs are well known and appreciated by the general public, in fact the vast majority of the malignancies induced in human beings by man-made radiation result from medical x-rays used in situations that nowadays we would consider unjustified. The irradiation of children who were supposed to have an enlarged thymus and the use of x-rays to epilate children with ringworm of the scalp are perhaps the two prime examples. It is difficult to make a detailed quantitative estimate, but it is certainly fair to say that for each Japanese survivor who contracted cancer as a result of the atom bombs, there must be at least 20 individuals who developed cancer because they were irradiated unjustifiably for medical purposes. This is a very sobering lesson, and one that we should keep in mind for the future. One wonders how many of the practices of medicine which are accepted today as par for the course will by some future generation be regarded with as much horror as we now regard the past use of x-rays in nonmalignant conditions, especially in children.

Extrapolating to Low Doses

We know more about the cancer-causing effects of ionizing radiation than about any other environmental carcinogen. There is a wealth of epidemiological information, summarized above, about irradiated human populations, not to mention the huge experimental literature about the effects of radiation on animals and their cells.

Yet the cancer risks from *low doses* of ionizing radiation remain a subject of bitter scientific dispute. There are two fundamental difficulties. First, epidemiologists cannot obtain clear direct evidence of cancer caused by radiation at the very low levels that interest government regulators, and are relevant to the exposure of large populations to tiny doses of radiation from nuclear power stations or even from the much larger doses of conventional diagnostic radiological procedures. The additional incidence of cancer is so small that the sample sizes necessary to obtain statistically convincing information are impractically large. For example, it would probably be necessary to study a sample of one hundred million women to detect clearly the additional cases of breast cancer caused by a single present-day x-ray examination that delivered a dose of .01 Gy (1 rad) to each breast. According to the best estimate, such a dose might induce six excess cancers in one million women a year, on top of the normal rate of about 1,910 cases in the same million women. Obviously, a sample of one hundred million women is not practical.

For this reason, it is necessary to approach the matter by observing much smaller populations exposed to far higher levels of radiation, such as the Japanese survivors of the atom bombs, or groups of patients receiving radiation for medical purposes as listed above. But then scientists come up against the second fundamental difficulty: they do not know how to extrapolate from a high dose to a low dose.

The simplest solution is to adopt a linear dose response model, which assumes that the risk is proportional to dose. Then, if a dose of 1 Gy (100 rad) causes, say, 600 excess cancers in a given population, then a dose of .01 Gy (1 rad) would produce 6 cases; that is, if the dose is reduced by a factor of 100, the number of cases is reduced by the same factor. It is doubtful whether any scientist really believes in his heart that the linear model can be correct. Both experimental evidence and theoretical considerations indicate that it may not be generally applicable. To begin with, linearity is not in accord with the experimental data available for animals. For x-rays, biological effect is

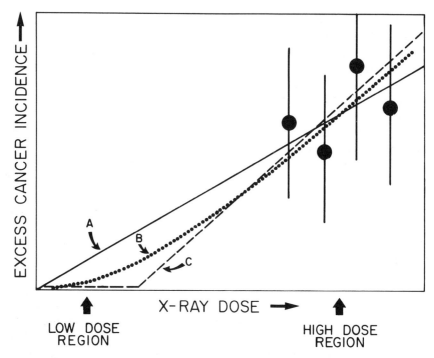

FIGURE 3.5. The problem of extrapolation. Most instances where radiation has been shown to produce an excess incidence of cancer in humans involve a few hundred cases exposed to large doses of radiation (several gray). The data are usually poor and give little idea of the curve shape. To extrapolate to the low dose region of interest, a model of some sort must be assumed. (a) is a linear extrapolation; simple, but probably an overestimate for x-rays. (b) represents a linear-quadratic relationship and is probably nearer the truth except that the shape is hard to know. (c) illustrates the threshold type of response—for which there is no evidence in the case of cancer. All models can be made to fit the high dose data, but result in quite different estimates of risk at low doses.

seldom proportional to dose. Smaller doses get progressively less and less efficient in producing an effect. However, while no one believes that the linear extrapolation is accurate, some consider it to be *prudent* because, hopefully, it will result in safety measures that are strict and have a built-in safety margin.

We know that if 10 people take 1,000 aspirins each, at least 9 of them will die. These are high dose data, comparable to the radiation data from the bomb. If now the same total number of aspirins (10,000) are given, one each to 10,000 people, what would happen? This is comparable to the peaceful uses of radiation, where large numbers of people get a very small dose from x-ray pictures, etc. A linear extrapolation, such as that proposed for radiation, would predict the same number of deaths, namely nine. We know that this is nonsense, because no one dies from taking one aspirin. If in spite of this the linear extrapolation were insisted upon as being "prudent," in the sense that it results in unnecessarily strict controls, then aspirins would be banned from routine use as a medication, because 9 deaths in 10,000 would be too high a price to pay for the relief of headaches. As a result many people would be denied the advantages of a useful drug. Being overcautious is not always prudent.

This analogy has a flaw in it, which illustrates an important point. In the case of a drug, such as aspirin, there is a clear threshold dose below which its use is absolutely safe. A linear extrapolation from high doses is patently seen to make no sense in this case. The situation is not so clear in the case of radiation. There is no dose of radiation which can be said to be perfectly safe. The lower the dose, the smaller the effect, but in no case can a safe threshold be *demonstrated* below which no effect is observed. There may be a threshold, but since we cannot prove its existence by experiment, it is wise to be cautious and assume that the smallest dose has the possibility of producing a deleterious effect.

At the other extreme from those who favor the linear dose response model are some radiologists and experimentalists who believe that experimental data generally fit a quadratic dose response curve. If so, risk would vary with the square of the dose, and the risk of low-level exposure would be far less than those indicated by the linear model. It is probable that the truth, as is often the case, lies somewhere between these two extremes.

Risk Estimates for Leukemia and Cancer

Despite the problems and uncertainties involved, several committees have reviewed the information concerning cancer induction in human populations exposed to relatively high doses of radiation and arrived at estimates of the risk involved when large numbers of people are exposed to much lower doses. The most credible of these are the committees set up by the U.S. National Academy of Sciences on the "Biological Effects of Ionizing Radiations" which reported in 1972 and 1981 (the so-called BEIR I and BEIR III reports) and the scientific committee of the United Nations (UNSCEAR) which reported in 1977.

Both the BEIR I and the UNSCEAR committees, with various degrees of reluctance, used the linear dose response model referred to above, while BEIR III considered other models as well and placed much credence on a compromise involving both linear and quadratic terms. Table 3.1 summarizes some of the most reliable risk estimates available for leukemia, thyroid cancer, and breast cancer, as well as for all types of cancer resulting from total body irradiation. These risk estimates result from the simple linear extrapolation, though in view of the uncertainty of the data the difference that results

Table 3.1. Risk Estimates for Cancer and Leukemia

	RISK ESTIMATES	
Type of Malignancy	Old Dose Units	S.I. Units
Leukemia	15 to 25 cases/million people exposed/rem	1.5 to 2.5 cases/1000 people exposed/Sievert
Thyroid Cancer	50 to 150 cases/million people exposed/rem	5 to 15 cases/1000 people exposed/Sievert
Breast Cancer	50 to 200 cases/million people exposed/rem	5 to 20 cases/1000 people exposed/Sievert
All Cancers from total body irradiation	1 death/10,000 people exposed/rem	1 death/100 people exposed/Sievert

if a linear quadratic model is used is of very little importance within the context of the present discussion.

The most reliable risk estimates available for leukemia are based on the information from the Japanese atom bomb survivors and from patients treated with x-rays for ankylosing spondylitis. The risk estimate is of the order of 1.5 to 2.5 cases of leukemia per thousand individuals exposed per sievert—or using the old dose units, 15 to 25 cases of leukemia per million individuals exposed per rem. It cannot be emphasized too strongly that this risk estimate is the slope of a linear extrapolation from the cases of leukemia produced in small numbers of individuals exposed to large doses. In the case of thyroid cancer the best risk estimate is 5 to 15 cases per thousand people exposed per sievert (50 to 150 cases per million per rem) and this is based on a study of children exposed to therapeutic doses of medical x-rays. The risk estimates for breast cancer is not very different from that for the thyroid, and amounts to between 5 and 20 cases per thousand people exposed per sievert (50 to 200 cases per million per rem), an estimate which relies to some extent on the data from the Japanese survivors, but more particularly on the experience in which female patients with tuberculosis were fluoroscoped with x-rays repeatedly. Perhaps the most widely used radiation risk estimate, quoted frequently on radio and television talk shows during the period of the Three Mile Island nuclear reactor accident, is the risk of cancer of all sorts from total body irradiation. This figure is somewhat uncertain since it can be based only upon data from the Japanese survivors; no other group of individuals has been exposed to total body irradiation and studied for so long. The risk estimate is of the order of 1 death from cancer per 100 people exposed per sievert (1 death per 10,000 per rem). It should be noted that in this case *deaths* from cancer are quoted whereas the previous risk estimates for leukemia, thyroid, and breast involved *cases* of cancer. Not all forms of cancer necessarily lead to death; this is true particularly of thyroid cancer which only leads to death in a small proportion of cases.

It cannot be repeated too often that these risk estimates are based on very tenuous data. Nevertheless, so long as caution is exercised, risk estimates do enable us to make "ball-park" estimates of the hazards which we face as a result of taming and using the peaceful atom, both in power production and in medical radiography. An attempt to balance risk versus benefit will be made in the last chapter of this book.

Implications of the Revised Dose Estimates for the Japanese A-Bomb Survivors

Since the appearance of the most recent UNSCEAR and BEIR reports on the biological effect of radiation there has been a further evaluation of the radiation doses received by the Japanese A-bomb survivors. At first sight it would seem to be remarkable that there is still debate now, 40 years after the bombs were dropped, concerning the doses of radiation involved. In fact, the dosimetry is a difficult problem. The weapons used on the two cities were quite different—one was a plutonium bomb, the other a uranium bomb. The weapon used at Hiroshima was nicknamed "Little Boy," the one at Nagasaki "Fat Man"—reflecting the different design, size, and shape. Weapons similar to the one used at Nagasaki were tested before and after at test sites in the desert, so some measurements are available of the radiation doses at various distances from the explosion. By contrast, the weapon used at Hiroshima was the only one of its kind ever used—none were tested beforehand and by the time a duplicate was assembled for measurements after the war, the nuclear test ban treaty had been signed, and it was never detonated. Consequently all of the dose estimates for Hiroshima are based entirely on calculations and computer models—there are no measurements available. It is ironic that the vast majority of excess cancer cases are at Hiroshima where the dosimetry is so uncertain.

The dose estimates previously used were called in question for a variety of reasons, in particular, the fact that it was a hot humid day (80% humidity, 26.7°C or 80°F) in Hiroshima when the bomb was dropped, which would lead to the absorption of much of the neutron radiation in the water vapor in the atmosphere. At all events, the new dose estimates are lower for gamma rays and particularly for neutrons. The final numbers are not yet available, and it may be several years before agreement is reached on new best estimates. The revisions are based on the most up-to-date weapons physics and represent the latest stage in a continuing process of development dating from the early 1950s.

The Japanese bomb survivors are not the only source of human information on the late effects of radiation, but they are the most important and figure prominently in any calculation of cancer risk estimates for radiation. Consequently, as the dose estimates are *decreased*, risk estimates for radiation-induced cancer must *increase* correspondingly. No doubt the UNSCEAR and BEIR reports will be revised and new cancer risk estimates made when the dosimetry revisions are complete. This is likely to be a lengthy and time-consuming process. A best first guess of competent experts in the field is that the risk estimates will go *up* by a factor of two to four. This best "guess" can be arrived at in two ways. First, by taking into account the changed dosimetry in Japan and accepting the preliminary figures; second, by deriving estimates from the (admittedly limited) human data from other sources—the so-called nonbomb data. A change in risk estimates by a factor two to four is really not very large in view of the great uncertainty in the estimates, which can never be regarded as better than ball-park figures. Consequently, when this book was written, it was decided to stick to the cancer risk estimates of the reputable organizations (essentially the UNSCEAR and BEIR reports) despite the fact that the dose revisions were receiving a great deal of publicity at the time, always bearing in mind that the estimates may be low by a factor of two to four.

2. Genetic Mutations that Affect Future Generations

If a mutation occurs in a germ cell, in either the sperm or the egg, the consequences will be felt, not by the individual in which the change takes place, but by some future generation.

The germ cells come from the gonads, a blanket term that covers both male and female sexual organs. The male sex glands, the testes, produce sperm; the female ovaries produce egg cells. A sperm and an egg cell unite to make one tiny organism, barely visible, but containing the thread of our inheritance. Both sperm and egg each contain 23 single chromosomes. When the two cells fuse, the 23 single chromosomes from the father's sperm pair with the 23 single chromosomes in the mother's egg to form the first cell of a new human being which contains 23 pairs of chromosomes, i.e., a total of 46.

Chromosomes and Genes

The chromosomes carry, in code form, all of the blueprints which specify a human being as opposed to a mouse or an elephant—the well-developed brain, the sensitive fingers, the upright stance. But over and above this, they contain the information to reproduce all of the characteristics which "run in the family." The offspring may have an aquiline nose like the father, or an oval face like the mother; he may take after grandfather by having a natural ear for music, or have ginger hair like his aunt. All of these characteristics, and hundreds more, are coded for in the 46 tiny chromosomes in the fertilized egg.

The chromosomes are long, thread-like structures made up of a complex material called DNA, which is itself a very long molecule. The backbone of DNA is made of molecules of sugar and phosphates, but these serve only as a framework to hold in place the important molecules which carry the code of inheritance. Attached to each sugar molecule is a chemical called a base; they come in four varieties—thymine, adenine, guanine and cytosine, usually known by their initial letters T, A, G, and C. This whole configuration is coiled tightly in a double helix. It is rather like a tiny spiral staircase; a chain of sugar molecules forms the rail on either side, bridged at regular intervals by pairs of nucleic acids which form the steps.

The order, or sequence, of the nucleic acids contains, in code form, the genetic information. Just as in Morse code, different combinations of dots and dashes can give all 26 letters of the alphabet, so in the DNA various combinations and various sequences of T, A, G, and C code for the production of the complex molecules of which living things are made. The basic idea of the code has been cracked, but we do not know many of the details. A section of a chromosome which carries, in code form, the information for a specific characteristic is called a gene. We speak of the gene for blue eyes, the gene for red hair, and so on. No one has ever seen a gene; like the atom, its existence, its nature, and the way in which it works, is inferred from cleverly designed experiments.

Mutations: Natural and Radiation Induced

Sometimes sections of the code may get mixed up; the sequence of base pairs may get altered. In this case the chromosome may contain a flaw, which will be passed on to all of the daughter cells produced by division. If a damaged gene or chromosome occurs in the sperm or egg, all the cells of the resulting embryo will repeat the flaw. If the embryo survives and eventually grows up to be a parent itself, the genetic fault will be passed to his children, and on through future generations. Any cell that contains chromosomes or genes that are abnormal, that have been changed in any way, is said to be a mutant cell.

A mutation that occurs in a somatic cell will affect only the individual during his lifetime. A mutation occurring in a germ cell is called a genetic mutation and will be passed on to succeeding generations. Radiation can cause breaks and changes in the DNA of germ cells and so produce an increase in the number of mutations over and above that which occurs spontaneously. The mutations produced by radiation are indistinguishable from those that occur naturally. Radiation does not produce new, unique, bizarre mutations—but rather increases the incidence to which the species is subject anyway.

The various types of mutations that occur naturally and that can be increased by radiation fall into the following categories:

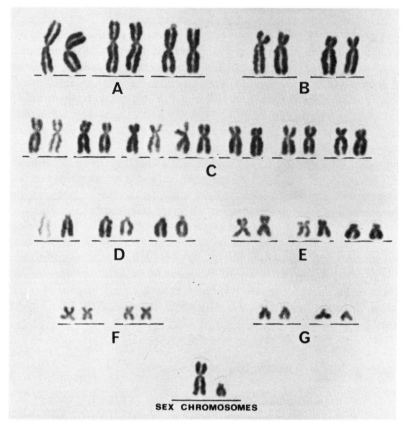

FIGURE 3.6. (a) Normal complement of human chromosomes, sorted out into the various groups.

1. Single gene mutations.
2. Wrong chromosome number—i.e., too many or too few, or chromosome aberrations, with the chromosomes rejoining incorrectly after they have been broken.
3. Frequent but mild mutations like those observed in *Drosophila* that cannot be identified as due to a specific identifiable or observable change in chromosomes.

Recessive and Dominant Genes

The fact that pairs of chromosomes contain corresponding sets of genes results in the possibility of recessive and dominant genes.

Eye color is the simplest example. The gene for blue eyes is recessive, and that for brown eyes dominant.

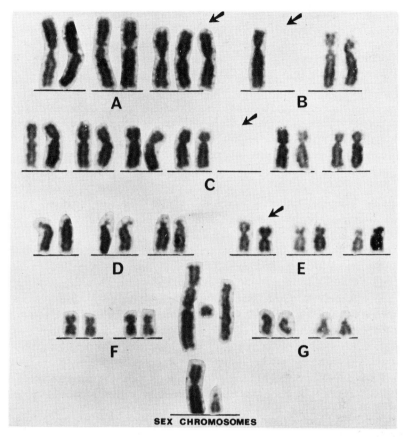

FIGURE 3.6. (b) Chromosome preparation from a radiologist who had worked with x-rays for many years. Note the many abnormalities. There is an extra chromosome in group A, while one is missing from group B and two from group C. Most spectacular are the aberrant chromosomes between F and G; two of them are dicentrics, i.e. have two centromeres. This is a result of incorrect joining of loose ends after the chromosomes have been broken by radiation. (Courtesy of Dr. Mary Ester Gaulden and W.B. Saunders Co.)

If a mother's pair of chromosomes both contain the gene for blue eyes, she must herself have blue eyes, and must hand the blue-eyed gene on to her offspring. Suppose the father is brown-eyed. This could be a result of both chromosomes containing the brown gene, or one brown and other blue, because brown is dominant. If both his chromosomes contain brown-eyed genes, his offspring must inevitably have brown eyes. But if the father's chromosomes contain one gene for brown eyes, and one for blue eyes, half of his sperm will contain genes for either eye color. Consequently, his offspring could have either blue or brown eyes. If he hands on his recessive blue-eyed gene to the offspring of the blue-eyed mother, the child will be blue eyed, because two recessive genes have met. Brown eyes always win, but blue eyes are a common recessive gene in a mixed race such as ours.

Two brown-eyed parents may both carry the recessive blue in one of their chromosome pairs. On average, the recessives will meet in one out of four children, while the other three will have brown eyes. The brown-eyed person who carries the recessive gene will, in outward appearance, be indistinguishable from the brown-eyed person who carries two brown-eyed genes. The existence of a recessive gene is impossible to detect until matched in the offspring with a similar recessive gene, when it comes out into the open at last. The variety of eye colors in our society probably reflects our mixed racial ancestry. Blue eyes and brown eyes add a little variety to the human scene, and either is perfectly acceptable from a functional and aesthetic point of view. They are normal variants.

Harmful mutations, however caused, may also be dominant or recessive. New mutations are constantly being created in the human population, due to radiation (natural or man-made), chemicals, or simply to mistakes occurring during cell division. Dominant mutations, by their very nature, show up immediately. They cause a variety of troubles. In fact there are about 500 known human diseases that can be attributed to a single dominant gene mutation, including polydactyly and retinoblastoma. The deformity that we recognize in most dwarfs is due to a dominant mutant gene. Since dwarfs rarely have children, the mutant gene responsible dies out after being expressed only once. By contrast, recessive genes are perpetuated in the population, and are a greater long-term problem. There are over 500 known human conditions attributable to recessive gene mutations, including Tay Sachs Syndrome, common in certain groups of Jewish people from Central Europe, and sickle cell anemia, common in blacks who originated in specific areas of Africa where this mutation, remarkably enough, confers some advantage in a situation where malaria is prevalent.

One type of diabetes, the failure of the pancreas to secrete insulin, is a disease caused by recessive genes. Mental defects, grave mental illness, many physical deformities, as well as a reduced resistance to disease can all be attributed to recessive genes.

Sex-Linked Mutations

There are some diseases that are inherited only by male children; these are said to be sex-linked. When a daughter is born, she inherits two sex chromosomes; an X-chromosome from her mother and an X-chromosome from her father. If there is a harmful recessive mutation in one of these chromosomes, there is a good chance that there will be a normal and dominant gene in the other, and the recessive will not be expressed. But when a son is conceived, the

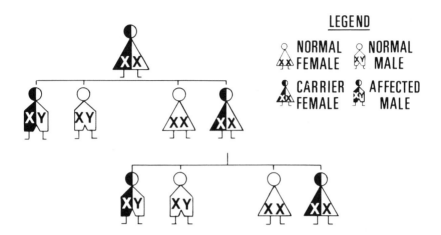

FIGURE 3.7. How hemophilia is inherited.

father donates a Y sex chromosome, which contains few of the genes present in the X sex chromosome from the mother. The sex chromosomes in the male (XY) represent the only instance where the pair of chromosomes do not contain parallel sets of genes. If the mother carries a recessive mutant gene on the single X chromosome that she supplies to her son, there may be no matching gene from the father. When this happens, there is no opposition to the mutant gene, and the defect or disease will be expressed.

By far the most common sex-linked defect is color blindness. About 1 in 12 men are color-blind, either partially or totally. It is very rare for a woman to suffer this disability because it could happen only when recessive mutants were present in both X chromosomes. Hemophilia is probably the most publicized sex-linked disease, which like color blindness, is essentially restricted to males. When a hemophiliac is cut, he bleeds copiously, because the blood lacks a vital clotting factor. It is a serious disability, and until recently victims of the disease seldom survived childhood.

When a woman has the recessive mutant in one of her X chromosomes, it will pass, on average, to half of her children. A son who receives the affected gene will inevitably be a hemophiliac. Daughters will be unaffected themselves, but on average, half carry the mutant gene and in that case will pass it on to half of their children.

Incorrect Chromosome Number

Genetic effects may be due to an incorrect number of chromosomes—either too many or too few. Down's Syndrome (the mongol child) is the best-known example and is due to an extra chromosome #21. Conversely, some rare forms of mental retardation are a consequence of one chromosome too few.

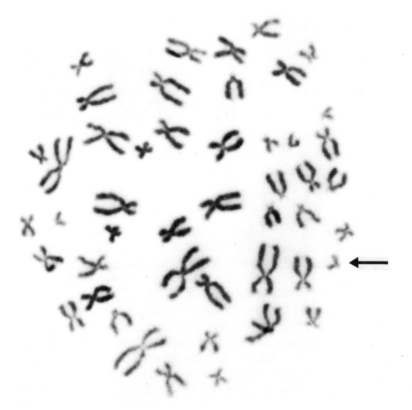

FIGURE 3.8. Illustrating how a small chromosome change can produce a serious biological effect. Shown is a chromosome preparation from a child with Down's Syndrome (commonly known as mongolism). This is a result of the single extra chromosome indicated by the arrow. (Courtesy of Drs. Julian Preston and Henry Luippold, Oak Ridge National Laboratory, USA.)

People who suffer from such severe disabilities seldom have children, and so these mutations die off at the same rate that they appear spontaneously in the population. Unlike gene mutations, which are much too subtle to be seen, some of these defects are so gross that they can be readily observed by a microscopic examination of the chromosomes of individual cells. In the fetus, chromosome breakage or rearrangement can occur spontaneously or after irradiation but usually results in embryonic death—but when it doesn't, it may produce gross physical abnormalities and/or mental retardation.

Frequent but Mild Mutations

An interesting genetic effect, much discussed of late, concerns mildly deleterious mutants, of no consequence to the individual but so frequent that the cumulative effects may be important. They can be observed only on a statistical basis in large populations of *Drosophila*, the fruit fly, from huge experiments on this system. The mean loss of viability due to frequent but mild mutations may be as much as 10 percent. Experiments indicate that many different small mutations may be involved, but overall the frequency may be 20 times greater than the frequency of lethal mutations. In *Drosophila*, therefore, they represent the biggest category of mutations, though there are data to suggest that these mild mutations make up a smaller fraction of the total for radiation-induced than for spontaneous mutations. We can only speculate on the possible importance of this in mammals in general, and humans in particular.

Summary of Mutations

Chromosome mutations of all types affect about one in a hundred of all children born alive. In addition, about five in a hundred are abnormal in some way due to gene mutations. To this must be added an unspecified number of lethal mutations that cause the embryo or fetus to be lost before birth.

The overwhelming majority of mutations must be considered to be harmful, or at best neutral. There are those who argue that some mutations are necessary for the upward evolution of the human species, but in fact the optimal mutation rate is probably zero, since the pressure of natural selection is greatly relaxed in a civilized liberal society and most children survive.

In the case of animals in the wild the law of the jungle is harsh, and any creature ill-equipped to fight the battle for survival is quickly eliminated. In human society things are more complicated. The victims of hereditary disease or deformity must often endure much suffering and great unhappiness through no fault of their own. In addition, they are a burden to the community. Hospital beds, medical care, and social services must all be provided to ease their lot. In a few instances, medical science has been able to help the victims of recessive mutations, diabetics for example, to overcome their handicap and live normal and healthy lives. But this is not often the case. We cannot renovate defective genes, or undo a mutation. A partial answer to this dilemma is a further development and wider use of genetic counseling. An increase in the use of any agent which causes mutations, including radiation, must be viewed with some alarm.

GENETIC EFFECTS OF RADIATION

Essentially all we know about radiation-induced mutations comes from studies in the laboratory; there is no direct evidence in the human of mutations produced by radiation. The largest group of humans available for study are the descendants of the Japanese exposed to the atomic bombs at Nagasaki and Hiroshima. Up to the present time there is no evidence of any genetic changes in the children born to these individuals. However, the number involved (about 110,000) was small by genetic standards, and insufficient time has passed for all recessive mutations to show up, since several generations must elapse before they are likely to be expressed. Consequently, *absence of evidence should not be construed as evidence of absence*! A genetic effect large enough to be observed in a sample of this size would not be expected. In the absence of data for man, the best that can be done is to assume that the results obtained with laboratory animals apply to man.

Up until the Second World War, most radiation genetic experiments were performed with the fruit fly *Drosophila*. This tiny creature breeds at a prodigious rate, and thus large numbers and many generations may be studied in a short period of time. In *Drosophila*, it is not difficult to see the effect of lethal mutations in terms of a reduction of the number of offspring, or the appearance of recessive mutations several generations later which result in obvious deformities of the wing or changes in eye color. Doses of thousands of rads are needed to produce these genetic effects because insects tend to be much more resistant to radiation than humans. A number of important conclusions were drawn from these *Drosophila* experiments, which were used to shape public opinion about the genetic effects of radiation up until the 1950s. First, the estimate of the dose required to double the spontaneous mutation rate, based on the *Drosophila*, was as low as 5 rem. Second, it was found in *Drosophila* that the mutation incidence was the same for a given dose of radiation whether the radiation was delivered in a single prompt exposure or whether it was spread out over a period of time. This led to the worrisome idea that radiation effects were cumulative, that a little radiation given today, a little radiation next week, a little radiation next month, would produce genetic effects which would all add up and become a permanent part of the genetic load carried by the population exposed.

The Megamouse Project

It was clearly an undesirable situation for policies concerning radiation protection to be based solely on experiments with the fruit fly *Drosophila*, but since about a million animals are required for the most modest genetic experiment, anything larger than *Drosophila* could not be contemplated until the years following the Second World War, when a great deal of money became available for radiation research as a result of the use of atomic weapons at Hiroshima and Nagasaki. At that time, large-scale experiments were planned using the mouse. Two different approaches have been used to study the genetic effects of radiation in this species, which now form a solid data base for an estimation of the hazards associated with the widespread use of man-made radiation. The first of these studies, performed at Oak Ridge

National Laboratory in the United States, is commonly known as the "mega-mouse" project because it was planned to use more than a million mice; in fact, the study eventually involved the use of over seven million mice. The strain of mice chosen is characterized by seven specific locus mutations which occur spontaneously in this strain, six of them involving coat color changes and the seventh involving a shortened ear. Examples of the coat color changes are shown in Figure 3.9. When animals are exposed to radiation, the inci-

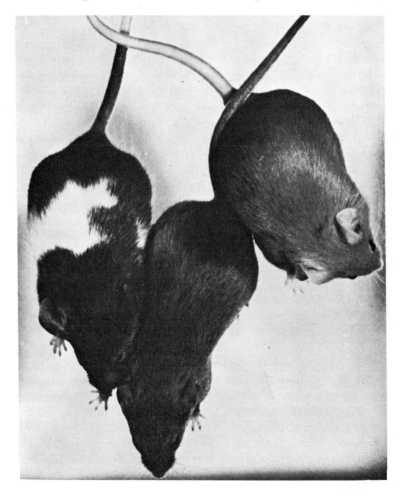

FIGURE 3.9. Illustrating coat color changes in the mouse, which are a result of mutations produced by irradiation of one of the parents before conception. These are three of the seven readily recognized mutations used to study the genetic effects of radiations at the Oak Ridge National Laboratory, U.S.A. (Photograph by courtesy of Dr. W.L. Russell, Oak Ridge National Laboratory.)

dence of these seven mutations is increased over and above the level that occurs spontaneously. The results of these costly and time-consuming experiments turned out to be very complicated indeed, but they may be summarized as follows: (a) The radiation sensitivity of the different mutations studied varies significantly, by a factor as large as 20, so that any discussion of sensitivity must involve an average value; (b) in the mouse there turned out to be a substantial dose rate effect, so that spreading the dose of radiation over a period of time greatly reduced the number of mutations produced by a given dose. This conclusion was in direct contrast to the situation for *Drosophila* and removes the nagging doubt that genetic effects of radiation are accumulative over a long period of time; (c) the male was more sensitive than the female to radiation-induced mutations. Indeed, at low dose rate, no significant increase in mutation rate was observed in the female following irradiation at low dose rate, even for doses approaching the lethal level. At low dose rates, radiation appears to be a very weak mutagen, particularly in the female; (d) the number of mutations produced by a given dose of radiation decreases significantly if a time period is allowed to elapse between irradiation and conception. This was first noticed in the male, and could be correlated with the stage in the development of the sperm at which the radiation was delivered, but in later experiments the same trend was found to apply to the female. This already constitutes the basis for practical genetic counselling; if a person is exposed to a substantial dose of radiation, either accidentally or as a consequence of medical x-ray procedures that involve exposure of the gonads (the sex organs), then it is recommended that a planned conception be delayed for a period of about six months in order to minimize the genetic impact of the radiation; and (e) based on the "megamouse" project, it is estimated that the doubling dose, that is the dose of radiation required to double the natural or spontaneous rate of mutation, is in the range 0.5 to 25 sievert (50 to 250 rem).

These experiments with the mouse contrast sharply with the earlier *Drosophila* work, and have lessened our concern with the genetic effects of radiation. There are two principal reasons: First, the doubling dose appears to be much higher than was previously thought. Second, there appears to be a substantial dose rate effect, so that if radiation is spread out over a period of time the genetic effects are minimized. In a practical situation when human beings receive radiation, it is usually spread out over a period of time.

Genetic Impact of Radiation in the First Generation

A second and totally independent series of experiments that allows an estimate of the genetic risk of radiation was also performed at the Oak Ridge National Laboratory and involved a particular strain of mice in which the frequency of 37 different skeletal anomalies following irradiation was measured in the first generation offspring following the irradiation of their parents. Most of the anomalies to the skeleton are very slight and can only be detected by a trained eye. This information is transferred to the human by allowing for the fact that only 10 percent of the genetic anomalies in humans involve the skeleton, compared with the total spectrum of anomalies. This technique allows an estimation to be made of the impact of radiation in the first generation.

SUMMARY OF GENETIC EFFECTS OF RADIATION

The principal facts we know concerning the genetic effects of radiation, deduced from these two studies in the mouse, can be summarized as follows:

1. Only one to six percent of spontaneous or natural mutations in the human can be ascribed to naturally occurring background radiation. In the past it had been suggested that spontaneous mutations may be due to the low levels of natural background to which all life is exposed; now that we know the sensitivity of radiation-induced mutations, it is clear that this simply is not the case.
2. Humans are not more sensitive than mice to the induction of genetic mutations by radiation, based on the lack of an observable genetic effect in the descendants of survivors of Hiroshima and Nagasaki. As previously mentioned, there is no evidence of any genetic effect in the children of survivors irradiated at Hiroshima and Nagasaki, nor would an effect be expected in such a comparatively small number if the sensitivity of humans is similar to that of the mouse; if humans were vastly more sensitive than the mouse then an effect would be expected in the 100,000 people exposed, but this is not seen.
3. The doubling dose, that is the dose required to double the natural or spontaneous level of mutations, is in the range of 0.5 to 2.5 sievert (50 to 250 rem) based on the specific locus mutations in the mouse.
4. Based on the same specific locus mutation data, it is estimated that a dose of 0.01 Sievert (1 rem) delivered to each generation of a population, in perpetuity, would increase the spontaneous mutation rate by about 1 percent. That is, if a population is exposed to this dose each generation for many generations, then the radiation might induce 60 to 1,100 mutations per million liveborn compared with the natural or spontaneous level of about 107,000 deformities per million.
5. The direct estimate of the genetic effects in the first generation following irradiation is 500 to 6,500 additional disorders per million liveborn per sievert of parental irradiation (5 to 65 per rem). This is based on the system of observing deformities in the skeleton.

The extensive, expensive and time-consuming experiments in the mouse over the past 30 years have led to a downgrading of concern regarding the genetic hazards of ionizing radiation. In the 1940s and 1950s, the genetic effects of radiation were considered to be the most serious potential hazard of exposing populations to low doses of radiation, but over a period of time concern for the genetic effects have been replaced by concern for somatic effects such as the production of leukemia or cancer. The principal reason for this change of emphasis or concern is that mouse data indicate that radiation appears to be a weak mutagen in mammals. The dose of radiation required to double the natural incidence of mutations is in the range 0.5 to 2.5 sievert (50 to 250 rem) compared with the previous studies with *Drosophila* which indicated that it might be as low as 0.05 Sievert (5 rem), coupled with the observation that there is a substantial dose rate effect in the mouse, so that radiation spread out over a period of time produces much fewer mutations than the same dose given as an acute exposure. Nevertheless, the impact of the genetic effects of radiation must be considered in assessing the balance between risk and benefit; an attempt to do this will be made in the last chapter.

3. Effects on the Embryo and Fetus as a Consequence of Irradiation During Pregnancy

There are two quite different reasons for defective development of the embryo and fetus. First, there are hereditary or genetic effects, already discussed. Second, there are external influences including exposure to ionizing radiations during embryonic or fetal development; defects due to such causes are known as teratogenic effects. (From the Greek, *Teras*, a monster.) The latter reason is the topic of this section.

It is just not possible to learn anything by checking the records of women who had x-ray pictures when pregnant, to see if they subsequently delivered abnormal babies. The sad fact of life is that, even without radiation, one in every twenty babies born has a defect or anomaly of one kind or another. A slight increase in this number, due to a very small dose of radiation, would be hopelessly difficult to detect in the human population because when a defective child is born to an irradiated mother, there is no way of knowing whether the anomaly was due to the radiation, or whether it would have occurred anyway. Radiation does not produce a special kind of defect which can be recognized, it simply increases the chance for the fetus exposed of being one of the unfortunates which develops in an anomalous way.

Because of the impossibility of studying the effects of small doses on human embryos and fetuses, experiments have been performed in which large numbers of small animals, such as mice or rats, have been irradiated under carefully controlled conditions. The conclusions from such experiments must be assumed to apply to the human. While caution must always be exercised in

FIGURE 3.10. (Top) Newly fertilized mouse egg. (Center) During preimplantation, the embryo consists of a limited number of cells. For example by the third day, the mouse embryo contains about 16 cells. (Bottom) By about 3½ days postconception in the mouse, which corresponds to 9 or 10 days in the human, the embryo becomes embedded in the wall of the uterus and about this time cells begin to differentiate to form specific tissues and organs. (Courtesy of Dr. Peterson, University of California at San Francisco.)

extrapolating from mouse to man, this is the best we can do at the present time.

Most of what we know, then, about the effects of radiation on the developing embryo and fetus comes from laboratory experiments with rats and mice. Some important human data come from the study of the offspring of Japanese survivors who were pregnant at the time of the A-bomb attacks on Hiroshima and Nagasaki.

Preimplantation

The development of the unborn child can be divided into three major time periods. After conception, the fertilized egg undergoes rapid division until about the ninth day, when it becomes firmly embedded in the wall of the uterus; this period is known as *preimplantation*. A sizeable dose of radiation delivered during this period may kill the newly formed embryo, which at this stage consists of only a limited number of cells, but does not appear to produce other effects such as growth retardation or gross anomalies charac- teristic of later times. In other words, radiation appears to have an all or

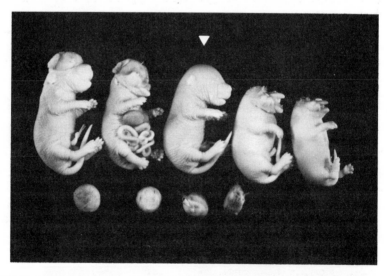

FIGURE 3.11. Litter from a mouse irradiated with x-rays while pregnant. At the bottom of the picture are four embryos that were resorbed, i.e., killed by the x-rays. Of the five fetuses that were alive, the center one (marked by an arrow) is normal. The other four show gross anomalies, either of brain formation, or development of the intestines or general stunting of growth. This, of course, is an extreme case produced in an experimental animal by a large dose of x-rays. (Picture by courtesy of Dr. Roberts Rugh.)

nothing effect at this stage of development. The death of so early an embryo, which occurs frequently from natural causes, too, would probably go unnoticed in most cases.

Organogenesis

The period of development from the 9th day to the 6th week after conception is known as *organogenesis*, because it is then that cells begin to differentiate to form the specialized organs and parts of the body—the eyes, the brain, the arms and the legs. While organs and limbs are being formed, the embryo is most vulnerable to disease, to drugs, and also to radiation.

It was during this period that the drug thalidomide, given to pregnant women in Europe to suppress morning sickness, resulted in babies with stunted arms and legs. During this period, an attack of Rubella, normally a trivial infectious disease, can produce gross abnormalities in the fetus. Modest amounts of radiation can also have catastrophic consequences during this time. The whole spectrum of defects can occur, including cleft palate, stunting of the limbs, an abnormally developed brain, and so on. During the period when the organs and limbs are developing, they do so in a strict sequence; the radiation will affect those that happen to be developing at the moment of exposure.

As well as gross anomalies, irradiation early in organogenesis can lead to growth retardation, which shows up as an offspring of reduced size at birth. Neonatal death can also result from irradiation at this time, i.e., death at or about the time of birth. In terms of human misery, however, the live birth of a grossly deformed child is perhaps the biggest disaster, far worse than the death of the embryo, which is serious enough of itself.

Fetal Period

The period of gestation from the 6th week to term is known as the fetal period. Little information is available from animal experiments for irradiation at this time, except that large doses produce permanent growth retardation—i.e., the offspring are smaller than usual at birth and remain below average in size through their adult life.

Children of Survivors of Hiroshima and Nagasaki

In the offspring of the Japanese survivors, a reduced head diameter and mental retardation figure prominently as defects observed. These effects had not been reported from the studies with rats and mice, but then such effects would be difficult to observe in small animals. How do you assess the IQ of a mouse? An additional factor unique to the human is that the development of

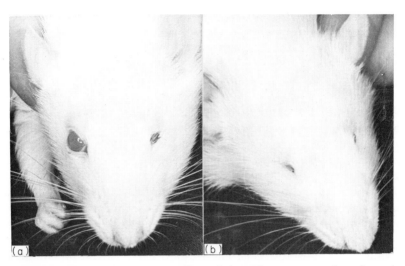

FIGURE 3.12. Two rats from the same litter; their mother was exposed to a large dose (100 R) of x-rays while pregnant. The rat in (a) has a normal right eye, but the other is defective. Both eyes are defective in the rat in (b). (Courtesy of Dr. Roberts Rugh and the *Journal of Military Medicine.*)

the central nervous system extends over a considerable fraction of the total time in utero. Effects on the brain and central nervous system were seen for irradiation in organogenesis on the fetal period.

Of the Japanese bomb survivors irradiated in utero, the number of cases of severe mental retardation above that which is expected is about 14. More cases occurred at Hiroshima than at Nagasaki, though there is no plausible reason for the difference. Cases did occur at Hiroshima in the dose range of 0.1 to 0.5 Gray (10 to 50 rads), though the doses are at present under revision. The obvious mechanism for mental defects and other gross organ defects is cell killing, since the developing embryo and its organs consist largely of relatively small numbers of dividing and differentiating cells. On general biological grounds, one would expect a threshold type of response—i.e., no deleterious effect up to a certain dose, and this is roughly what is observed in the Japanese survivors, though the numbers are small.

The 10-Day Rule

A most serious practical problem occurs from time to time when a woman receives a whole series of x-ray films involving the stomach or pelvis and subsequently discovers that she is pregnant. To make matters worse, as likely as not, the dose will have been received during the early weeks after conception, because it is only during this early stage that a pregnancy may remain undiscovered. Unfortunately, this is the very period when the most disastrous consequences result from a given amount of x-rays.

The only completely satisfactory solution to this problem is to ensure that it never happens in the first place. This may be achieved by ensuring that women of childbearing age receive x-rays of the stomach or pelvis only during the first 10 days after the start of a menstrual period, i.e., when it is quite certain that they are not pregnant. This is called the 10-day rule, and causes some problems of organization and scheduling. Nevertheless it has been introduced into hospitals in some countries, particularly Great Britain where medicine is organized under the National Health Service. In practice the rule can only be applied to "elective" procedures that are not urgent, such as a chest x-ray or x-ray of the lower back before a new job or to monitor routinely the course of a disease. X-rays needed in an emergency cannot be delayed for weeks or months—those that can should never have been ordered in the first place. Nevertheless it is prudent for a woman of reproductive age to tell the physician if there is any possibility that she is pregnant before a x-ray procedure is performed.

Accidental Exposure of the Embryo

Despite the best-laid plans, there will always be a few cases when, because of clinical urgency or unusual accident, an early developing embryo will be exposed to radiation. There are times when, after an auto accident for example, urgent and extensive x-ray examinations of the pelvis are called for as a matter of life and death, and it is out of the question to book ahead to the start of a new menstrual cycle. What to do in these cases? The first step is to make a careful estimate of the dose received by the embryo. No dose level can be regarded as completely safe.

At the same time it is impossible to attribute a given anomaly in a baby to a small dose of radiation received by the mother during pregnancy; all that can be said is that radiation increases the chance of an anomaly, and that this chance goes up with dose. Many experts believe that there is enough evidence in the human to recommend that if a dose in excess of 0.1 Gray (10 rads) is received by the embryo during the first six weeks after conception, then an abortion should be performed because of the strong possibility of producing an abnormal child. This figure is written into the law in some Scandinavian countries. It is clearly flexible, and may vary with other circumstances. For example, if a woman wants a baby, and for medical reasons another pregnancy is not possible, then she may choose to take the risk and have the baby despite a dose of 0.1 Gray (10 rads). Alternatively, if a mother-to-be does not want the baby in the first place, or if she is young and fertile and likely to enjoy many more pregnancies, then there is little point in taking the risk of giving birth to a deformed child, and an abortion may be justified after a dose of less than 0.1 Gray (10 rads). Whatever happens, the risks should be explained to the mother-to-be, who must clearly be deeply involved in the final decision.

4. Immediate Death Due to Large Radiation Doses

There are two words in the title that need to be explained. First, "immediate" death means death within a matter of hours, days, or weeks, the exact time scale being related directly to the size of the radiation dose; this is in contrast to the time scale involved with the late effects already discussed which take many years to develop. Second, a "large" dose of radiation means a single prompt exposure to hundreds or thousands of rads.

Immediate radiation death has been extensively studied with many different types of animals and is reasonably well understood. The sensitivity of different species varies considerably, with small rodents such as rats and mice being relatively resistant in general, while larger animals such as dogs, sheep or pigs appear to be more sensitive to radiation killing. Cold blooded animals such as reptiles are usually more resistant than mammals, and insects are much more resistant by a large factor. In the human, experience is limited to the Japanese exposed during the atom bomb attacks on Hiroshima and Nagasaki, and to a handful of people irradiated in accidents at nuclear installations. It is irrelevant to a discussion of the effects of medical exposure or irradiation of the public by routine releases of radioactivity from nuclear power reactors, where the doses involved are a hundred to a thousand times too small to cause immediate death.

The nuclear industry has an incredible safety record in terms of the number of workers killed or injured. One of the consequences of the impressive safety record of the nuclear industry is that our knowledge of radiation death in the human is very limited, and this is one instance where we are happy that our information remains incomplete. Generating electricity by nuclear power is certainly cheaper in terms of human life and suffering than mining coal, which kills or maims so many men. It might be noted at this point that over 100,000 coal miners have been killed in the United States during this century.

Following whole-body exposure to large doses of radiation, the mode of the death and the length of the survival time are dependent on the magnitude of the dose. Three distinct modes of death can be identified, although in actual accidental exposures a great deal of overlap is frequently seen. At very high doses, in excess of 100 Sievert (10,000 rem), death occurs in a matter of hours and appears to result from damage to the central nervous system or to vascular breakdown; this mechanism of death has come to be known as the central nervous system (CNS) syndrome or the cerebral-vascular syndrome. At intermediate dose levels on the order of 10 Sievert (1000 rem), death occurs in a matter of days and is associated with destruction of the lining of the gut. This mode of death is knows as the gastrointestinal syndrome. At lower dose levels, of the order of 2 to 5 Sievert (200 to 500 rem), death occurs several weeks after exposure due to effects on the blood forming organs; this mode of death has come to be known as bone-marrow death or the hemato-poietic syndrome syndrome. The exact cause of death in the CNS syndrome is by no means clear; by contrast in the case of both the other modes of death, namely, gastrointestinal and the bone-marrow syndromes, death is due to depletion of the cells of critical self-renewal tissue, the lining of the gut or the blood forming organs respectively. The difference in the dose levels at which these two forms of death occur and the differences in the time scales involved reflect variations in the populations of cells in the two self-renewal systems and in the amounts of damage that can be tolerated in the different tissues before death ensues.

The Central Nervous System Syndrome or Cerebral Vascular Syndrome

A total body dose on the order of 100 Sieverts (10,000 rem) results in death in a matter of hours. While death in this dose region is due principally to failure of the central nervous system, all other organ systems will also be seriously damaged. Of course the gastrointestinal and the hematopoietic systems will both be severely damaged, too, and will ultimately fail just as they do at lower doses. But the failure of the central nervous sytems brings death so quickly that the consequences of the failure of the other systems do not have time to express themselves. The symptoms which are observed may be summarized briefly as follows. There is a development of severe nausea and vomiting, usually within a matter of minutes. This is followed by disorientation, loss of coordination of muscular movement, difficulty with breathing, diarrhea, convulsive seizures, coma, and finally death. There are two or three instances of accidental human exposures that resulted in doses high enough to produce this type of death; one such case will be described briefly: In 1964 a 38-year-old man, working in a uranium recovery plant, was involved in an accidental nuclear excursion. He received a total-body dose estimated to be 88 Gray (8800 rad) of which 22 Gray (2200 rad) was due to neutrons and the rest due to γ-rays. He recalled seeing a flash and was hurled backwards and stunned; however, he did not lose consciousness and was able to run from the scene of the accident to another building 200 yards away. Almost at once he complained of abdominal cramps and headache, vomited, and was incontinent of diarrheal stools which were bloody. The next day the patient was comfortable, but restless. On the second day his condition deteriorated; he was restless, fatigued, apprehensive, short of breath, and had greatly impaired vision, and blood pressure could only be maintained with great difficulty. Six hours before his death he became disoriented and blood pressure could not be maintained; he died 49 hours after the accident.

At these dose levels the exact and immediate cause of death is not fully understood. While death is usually attributed to events taking place in the central nervous system, much higher doses are required to produce death if the head alone is irradiated rather than the entire body; this would suggest that effects on the rest of the body are by no means negligible. It has been suggested that the immediate cuase of death may be an increase in the fluid content of the brain, due to leakage from small vessels, resulting in a build-up of pressure within the bony confines of the skull.

The Gastrointestinal Syndrome

A total body exposure of about 10 Sievert (1000 rem) commonly leads to symptoms characteristic of gastrointestinal death: nausea, vomiting and prolonged diarrhea. Individuals lose their appetites and appear sluggish and lethargic. Prolonged diarrhea, extending for several days, is usually regarded as a bad sign because it indicates that the dose received has been more than a 10 Sievert (1000 rem) and will inevitably prove to be fatal. After a few days the individual shows signs of dehydration, loss of weight, emaciation and complete exhaustion; death usually occurs after a few days, the exact time varying with the species involved. The symptoms that appear and the death which follows are attributed to a denuding of the lining of the gut, which is a classical example of a self-renewing structure. The surface of the gut is composed of a vast number of villi, finger-like protrusions composed of cells that are fully differentiated, able to perform the special functions of absorbing nourishment from the food, but incapable of further division. These functioning cells are continually worn away and sloughed off from the tips of the villi as food passes through the digestive system, and must be continually replaced by cells born in the crypts, which are factories of actively dividing cells located at the base of each villus. When the gut is exposed to a large dose of radiation, the Achilles heel of this self-renewal tissue is the population of dividing cells in the crypts. The radiation sterilizes a proportion of these cells, prevents them from dividing, and thus cuts off the supply of new differentiated cells to replace those worn off from the tops of the villi. The effect of the radiation is not apparent immediately, because the differentiated functioning cells of the villi are unaffected by the radiation and continue to perform their specialized functions. However, in the course of time, as the villi are worn away, there are no new cell replacements. As a result the gut is denuded of villi, and gross holes appear. Death soon follows as a result of loss of body fluids, and because of gross infections since the bacteria of the bowel, harmless and even essential in their own place, escape into the blood stream. The time delay between irradiation and the moment of crisis reflects the lifetime of the mature differentiated cells which make up the villi. In the case of small rodents, such as mice, this time period is about three days.

There is probably only one example known of a human suffering a gastrointestinal death. In 1946, a 32-year-old white male was admitted to the hospital within one hour of a radiation accident in which he received a whole-body dose of neutrons and γ-rays. The doses were very uncertain in this early accident, with estimates of total-body exposure ranging from 11 to 20 Gray (1,100 to 2,000 rad). In addition the man's hands received an enormous dose, possibly as much as 300 Gray (30,000 rad). The patient vomited several times within the first few hours of the exposure. His general condition remained relatively good until the sixth day, when signs developed of severe bowel obstruction. On the seventh day, liquid stools containing blood were noted. The patient developed signs of circulatory collapse and died on the ninth day postirradiation. At the time of death, jaundice and spontaneous hemorrhages were observed for the first time. At autopsy, the lining of the gut was found to be denuded and blood cultures showed the presence of *E coli* bacteria.

Death Due to a Failure of the Blood

At doses of a few Sievert (few hundred rem) death, if it occurs, is a result of radiation damage to a different self-renewal tissue, namely the blood forming organs. The bone marrow produces the red cells which circulate in the blood for several months and carry oxygen to every cell in the body. The bone marrow or lymph nodes also produce the platelets, which have a very short and useful life, and contain the blood clotting factor. A simple cut will bleed for a very long time in persons who have a platelet dificiency. If the platelets and certain chemical substances are present in the right amounts, the blood around the cut grows steadily thicker until it clots and seals the leak.

The lymph nodes and the thymus gland make only white blood cells. These constitute the security guards against invasion of the body by foreign objects. If a break in the skin allows alien bacteria to enter into the tissues and cause an infection, white blood cells rush to the spot and try to surround the invader. Painful swelling may result, but usually the infection is localized in this way. If the infection is a general one, as in the case of influenza or a more serious illness such as smallpox, invasion may be temporarily successful, but eventually it is the white cells that overcome the virus and develop a life long immunity in our bodies from further attacks by that particular agent.

The white cells provide immune response, the red cells carry oxygen to the cells of the body, the platelets prevent us from bleeding to death from the simplest cut. It is essential to health, therefore, not only to have cells of the right type, but have them in the right proportions. A dose of about 4 Gray (400 rads) has little or no effect on the circulating blood cells, which are mature functioning cells incapable of further divisions, but kills some of the primitive stem cells in the bone marrow, the lymph nodes and the spleen, depriving the body of the dividing cells which produce future replacements for the circulating blood. Apart from a brief period of nausea, which clears up after a few days, no effects of the radiation at this level are apparent for about a month, by which time the circulating cells reach the end of their useful life, die, and are removed.

Since many of the stem cells have been killed by a dose of a few Gray (few hundred rads), there are no replacement blood cells available. The exposed person suffers from hemorrhages due to the lack of platelets, tiredness and weakness because of anemia (lack of red cells), and at the same time resistance to infection is at a minimum because of the sparsity of white cells. If the dose is large enough, the person will die; if the dose is a little smaller, the person may recover after a period of crisis. Although there have been relatively few examples of humans exposed to a radiation dose of this magnitude, it is estimated that the dose that would kill half of a human population of young healthy adults with no medical intervention is about 3.25 Gray (325 rads). Within a given population of human beings, there are many factors that influence the response of the individual to whole-body irradiation. For example, the very young and the very old appear to be more radiosensitive than the middle-aged or the young adult. The female, in general, appears to have a greater tolerance to radiation than does the male.

Victims doomed to die can be saved by appropriate treatment if the dose is not too high. Transfusions of platelets and red cells can temporarily prevent bleeding and anemia. Careful nursing, isolation, and antibiotics can avoid the complications of an infection. By such means, an irradiated person can be helped over the period of crisis until his own body can begin to make the required circulating blood cells again. Some individuals who might have died can be saved. In some countries, notably Germany, an "isolation" technique is recommended. The victim of a radiation accident would be "sterilized" externally by repeated bathing in antiseptic solutions and then given a large dose of antibiotics. He would then be isolated in a airtight plastic unit and fed sterilized food so that he does not come into contact with bacteria or other pathogens in the environment during the period that his blood elements are depressed. Such techniques have not been put to the test because of the sparsity of radiation accidents.

In the late 1950s, a group of Yugoslav scientists were accidently exposed to doses of radiation estimated to be up to 9 Gray (900 rads). At this dose level, none would be expected to survive, so that extreme measures were called for. Six of the victims were sent to Paris and received massive bone marrow transplants; one died early of hepatitis, probably as a result of the transplants, while in the remainder the transplants were rejected, but the victims survived. By this time, a team of U.S. scientists had arrived on the scene to conduct an investigation, which concluded that the dose estimates were in error and that the victims had not received more than 3 to 4 Gray (300 to 400 rad), not a lethal dose with adequate medical care. Most experts conclude in retrospect that the Yugoslav victims survived in spite of the bone marrow transplants rather than because of them. This is born out by the fact that the one victim who did not receive a transplant survived to become the mother of four normal children. A few months after the Yugoslav accident, the famous Y12 accident occurred at Oak Ridge National Laboratory in Tennessee, in which five scientists were exposed to doses of 3 to 4 Gray (300 to 400 rads). They received no special treatment other than good nursing and tender loving care, but survived to a ripe old age.

CHEMICAL PROTECTORS AGAINST RADIATION

As early as 1949, it was observed that certain compounds administered prior to irradiation could protect mice from death by total-body exposure to x-rays. As described earlier, a total-body dose of about 4 Gray (400 rad) of x-rays leads to death by failure of the blood-forming organs in the human (30 to 60 days later), while a dose exceeding 10 Gray (1,000 rad) leads to death by 9 or 10 days due to damage to the gut. Radioprotective drugs raise the dose at which death occurs by a factor of up to two. This could be vitally important to an individual exposed to a dose of 4 to 10 Gray (400–1,000 rad)—it could make the difference between life and death. In addition, it modifies and reduces the early symptoms of nausea and vomiting produced by radiation.

The first compound found to be effective in mice was cysteine, but a number of other related compounds were soon identified that were at least as effective. It turned out that they all have in common a structure that includes a sulfhydryl (i.e., SH) group at the end of a long molecule. These compounds protect against the cell-killing effects of x- and gamma rays because they "scavenge," or mop up, the radiation-produced free radicals that are largely responsible for the biological effects produced by x-rays. Predictably, the military expressed a great interest in these compounds from the outset, and during the 1950s, the United States Army conducted an extensive program at the Walter Reed Army Hospital in Washington, D.C. to find or make better protectors. Hundreds of compounds were synthesized and thousands identified that work as protectors; a few are very much better than others! The aim was to produce a compound that soldiers could take under threat of a nuclear attack, to save their lives and preserve their fighting capability. The problem with the drugs initially discovered, including cysteine, and cysteamine, is that while they are efficient protectors, they are also very toxic. The thrust of the research in this field, therefore, was to find ways to reduce the toxicity of the drug while not compromising its efficiency as a radioprotector. Without going into any of the technical details, it is fair to say that this objective has been accomplished to some extent by the strategy of covering the sulfhydryl group on the compound with a phosphate group. The drug is activated within the body and becomes an efficient scavenger. One of the serious practical limitations of the drugs developed in this way is that they must be administered intravenously in order to be effective, rather than being given by mouth as a tablet, since the compounds are broken down in the acid in the stomach. This is a serious drawback for the practical use of these compounds, especially by the military in the field.

There are three interesting applications of these drugs. First, radioprotector tablets are carried routinely in the field pack of Russian troops stationed in Western Europe, to be taken in the event of a nuclear war. The compound involved is known as WR-638 in the United States (WR standing for Walter Reed) or CYSTAPHOS in the Soviet Union. The effect of these tablets would be largely psychological, because as already explained, these compounds need to be delivered by intravenous injection since they break down in the acid juices of the stomach when taken by mouth. The other limitation is that, while protectors work well against x- or γ-rays, they are largely ineffective against neutrons. Consequently, they would protect against the γ-rays from fallout but not from the initial neutron burst from a nuclear device.

Second, American astronauts carried radioprotector compounds with them on their flights to the moon, just in case a solar particle event caused a marked increase in the radiation levels in space. During the manned lunar missions, once the spaceship left orbit around the earth and headed for the moon in the so-called "Translunar Coast," the astronauts were then committed to a long mission. Their fuel supply was inadequate for them to turn around and return home prematurely, and they needed the gravitational field of the moon to slow down. They were committed, therefore, to go around the moon before they could return to earth and this involves at least 14 days. If there had been a major solar flare during the Translunar Coast the astronauts would have been inevitably exposed since they could not abort the mission. All of the manned space flights were timed to coincide with minimal solar flare activity, but it is never possible to guarantee that there will not be a solar flare. Accurate measurements have never been made of the doses in space during appreciable solar flare activity, but some estimates put the dose to the astronauts in the event of a major solar flare in the region of 4 Gray (400 rad), which approximates to the mean lethal dose. Even if the astronauts had not died, they would have been too ill to operate the spaceship and return it to earth. Consequently, the availability of a protector that would raise the mean lethal dose from 4 Gray (400 rad) to 8 Gray (800 rad) was a great comfort to the astronauts and might have been important. As it turned out, no manned space flight suffered a problem with solar flares and the maximum dose received by any astronaut so far in the United States space program has been only 50 milligray (5 rad).

Third, one of the compounds synthesized by the Walter Reed Army Hospital is currently being applied to the treatment of cancer by x-rays. This particular compound, WR-2721, is highly hydrophilic, i.e., it dissolves well in water but not in fats. Consequently, when it is given to a patient it quickly floods normal tissues but permeates into a malignant tumor more slowly. The basic strategy, therefore, is to give an infusion of this drug minutes before treatment with x-rays in order that the normal tissues may be protected by the drug while the tumor is not protected to the same extent. This is a neat idea and a simple strategy but it has not yet been proved to be effective.

In a wide variety of experimental systems, these sulfyhydryl compounds have been shown to effectively protect biological materials from the cell-killing effect of x- and gamma rays. It has not been similarly demonstrated that they can protect animals or human beings from the carcinogenic or mutagenic potential of radiation. However, a great deal of research effort is devoted at the present time to the study of a wide variety of antioxidants and sulfhydryl compounds as protectors against the production of cancer by both radiation and a variety of chemicals. Sulfhydryl compounds are found as natural constituents of all mammalian cells and may be one of the mechanisms that nature has provided to dispose of harmful free radicals produced by chemicals or radiation. It may be that in some instances nature can be helped by the addition of man-made compounds, particularly perhaps in high-risk groups of individuals.

4
What Comes Naturally

BACKGROUND RADIATION

Natural radiation is made up of cosmic rays which reach the earth from outer space, and radiation from the radioactive materials in the rocks of the earth and in the food which we eat. Cosmic rays are an external source of irradiation, since they arrive out of the skies and pass through our bodies. Naturally occurring radioactive materials give rise to both external and internal radiation of the human body. Radioactivity in the earth, and in building materials from which houses are made, give off rays which continually pass through our bodies; this constitutes an external source. At the same time, the food that we eat contains minute traces of radioactive elements which become incorporated into our bodies and constitute a continuous source of internal radiation. It has only recently been appreciated that breathing radon gas, given off from building materials, results in significant irradiation of the epithelial lining of the lung.

The study of natural background radiation is important because the human race has been continuously exposed to this radiation from the dawn of time, apparently with no deleterious effects. When we become concerned about the levels of man-made radiation to which we are exposed as a result of modern technology, it is extremely important to gauge these levels against the natural background which is the natural lot of man. This represents a long-term yardstick against which man-made innovations may be judged.

The radiation from natural sources remains relatively constant with time. It has been suggested that at certain remote times in the earth's development, the natural background radiation may have been many times higher than it is now, but in historical times, over the past few hundred years, it has not varied much in intensity. On the other hand, it does vary with location. The level of natural background radiation is 10 times higher in some parts of the earth than in others.

COSMIC RAYS

The intensity of cosmic rays arriving at the earth's surface varies with latitude and with altitude above sea level. It varies with latitude because the earth behaves like a giant magnet; cosmic rays are charged particles which are

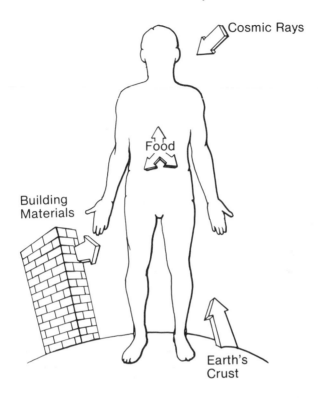

FIGURE 4.1. Illustrating the three major components of natural background radiation; cosmic rays from the sun or space, radioactivity in food, and radiation from the earth's crust, which in practice means from building materials since we spend so much time indoors.

deflected away from the equator and funneled into the polar regions. Consequently, cosmic-ray intensity is *least* in equatorial regions and rises towards the poles. In fact, the intensity remains relatively constant between 15° South, then rapidly increases to a latitude of about 50°, after which it remains practically constant again out to the poles. In addition there is a large variation of cosmic-ray intensity with altitude. At high elevations above sea level there is less atmosphere to absorb the cosmic rays, and so their intensity is greater.

Regions located near the equator and at sea level receive the minimum dose of cosmic rays, amounting to about 0.35 millisievert (35 millirems) per year. At sea level, but well away from the equator, at a latitude of about 50°, for example, the dose from cosmic rays amounts to about 0.50 millisievert (50 millirems) per year. This would correspond to cities such as London, New York, Tokyo, Toronto, Seattle or Moscow.

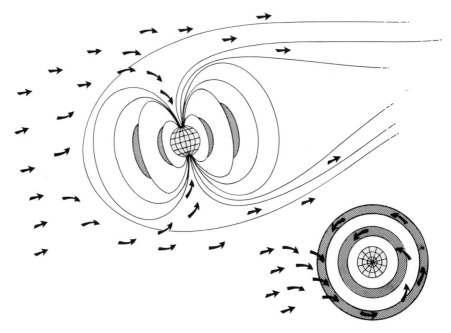

FIGURE 4.2. (Top) The earth is shown surrounded by the "lines" of its magnetic field. Showers of charged particles come from solar events on the sun and because of their charge are deflected by the earth's magnetic field; some miss the earth altogether, while others are funneled into the polar regions. This explains why the dose from cosmic radiation is low near the equator and high in the polar regions. It also explains the Aurora Borealis, or Northern Lights, caused by the intense cosmic ray particles coming into the poles. (Bottom) The earth viewed from above. Because of the spin of the earth, charged particles are trapped in layers by the lines of magnetic field. These are called the Van Allen belts. To leave the earth, a spaceship must pass through these layers.

Altitude is an even more important factor than latitude, and cities such as Denver, the "Mile High City," receive a dose from cosmic rays of approximately 0.9 millisievert (90 millirems) per year. The most elevated inhabited areas of the earth are at an altitude close to 15,000 feet, and here the cosmic-ray dose might rise to as much as 3 millisievert (300 millirems) per year. On the top of Mount Everest, the highest spot on the surface of the earth, the corresponding figure would be approximately 8 millisievert (800 millirems) per year.

In fact, most big cities, which house the majority of people on earth, are situated about halfway between the equator and the polar regions and close to sea level; there are no large cities near the poles, and none at really high altitudes. Because of this the average cosmic ray dose to the world population is about 0.5 millisievert (50 millirem) per year.

NATURAL RADIOACTIVITY IN THE EARTH'S CRUST

Naturally occurring radioactive materials are widely distributed throughout the earth's crust, and as a consequence man is exposed to the gamma rays emitted by them. The extent of the exposure involved varies from a value somewhat less than that due to cosmic rays in some places, to values many times higher than that due to cosmic rays in a few localities. In general, natural radionuclides are concentrated in granite rocks; limestones and sandstones are low in radioactivitiy, but certain shales are very radioactive, especially those containing organic matter.

In nature there are two very important radioactive materials, uranium-238 and thorium-232. These materials have a half-life of millions of years. As they decay and break up, not only do they emit radiation, but they also produce other radioactive materials with shorter half-lives. Uranium ore has been found in large quantities in Australia, Canada, Czechoslovakia, the Republic of the Congo, South Africa, the United States and the Soviet Union. Large deposits of monazite, the principal thorium-bearing mineral, are found in Brazil, China, India and the United States.

The average dose at a height of one yard above limestone is about 0.2 millisievert (20 millirems) per year, while for granite areas the corresponding figure is 1.5 millisievert (150 millirems) per year. These figures vary widely, however, and must only be regarded as approximate. In order to estimate the dose actually received by a human population in a given area, a carefully planned series of measurements must be made, indoors as well as out of doors, since most people spend the greater part of their lives inside a building of some sort. The dose rate inside a building depends mainly on the radioactive content of the material of which it is made. A Scotsman in Aberdeen, for example, living in a house built of granite, will be irradiated far more than a New Englander whose house is made almost entirely of wood.

Measurements have been made in 23 different areas within the United States of America, and the dose rate out of doors varies within the range 0.45 to 1.3 millisieverts (45 to 130 millirems) per year. Measurements made inside houses in the United States fall within the range 0.29 to 0.9 millisieverts (29 to 90 millirems) per year. Comparable measurements in other parts of the world are quite similar, except in a few special areas where the dose rate is very much higher due to the presence of large amounts of radioactivity in granite and in shale. For example, in large areas of France, in India and in Brazil, the dose-rates are several times higher than those quoted for the United States.

A source of natural radiation ignored until relatively recently is the dose to the surface of the lungs resulting from radiation from inhaled radon gas. This

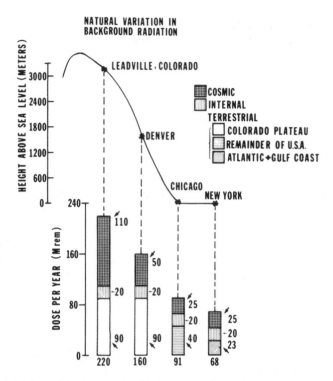

FIGURE 4.3. Natural background radiation comes from three sources: cosmic rays, radioactivity in the earth's crust, and radioactivity in the food that we eat. This figure shows the natural background radiation levels in representative cities in the United States. For each city the three components of the dose are quoted separately. For example, in New York City, 0.25 millisievert/year (25 millirem/year) comes from cosmic rays, 0.20 millisievert (20 millirem/year) from food and 0.23 millisievert (23 millirem/year) from radioactivity in the earth's crust. Doses are, of course, much higher in Colorado because a) cosmic ray levels are high because of the elevation and b) the rocks contain high levels of natural radioactive materials such as thorium and uranium.

arises from building materials and from the earth. Detailed national surveys have not been performed, but in most houses in the United States, the radon concentration appears to be about 37 millibecquerels/liter (1 pico curie/liter) in the living area, and much, much higher in the basement. The average dose to lung epithelium from radon and daughter products in the United States population is of the order of 5 millisievert (500 millirem) per year. This is the highest dose due to natural background. It is estimated that this component of natural background radiation may be responsible for 50 deaths per year from lung cancer in the United States. Perhaps the most disturbing feature is the extent to which radon concentration is increasing as people strive to seal their homes tighter and tighter to cut down heating bills. It is paradoxical to note that the radiation the average individual receives as a result of such conservation measures greatly exceeds the dose they would have received if nuclear reactors had been built to produce energy for heating in the United States.

NATURAL RADIOACTIVITY
WITHIN THE HUMAN BODY

Small traces of radioactive materials are normally present in the human body. These are ingested as a result of radioactive materials being present in tiny quantities in food supplies. Radioactive thorium, radium and lead can be detected in most people if very sensitive and extremely sophisticated techniques are used, but the dose-rates involved are very low indeed and very variable between one person and another. The usually quoted figure is certainly less than .01 millisievert (1 millirem) per year. The only isotope which makes a significant contribution to human exposure from ingestion is the radioactive isotope of potassium. Concentration of potassium in human beings varies considerably with age and with many other circumstances too. Several independent reports suggest that naturally occurring radioactive potassium results in a dose of about .02 millisievert (20 millirems) per year to the gonadal tissue, and cannot, therefore, be neglected as a cause of mutations in human beings.

HIGH NATURAL RADIATION AREAS

In a few areas of the world, the dose-rate from natural background radiation is considerably higher than that experienced by the majority of the human race. This high radiation background is due to the presence of larger than normal amounts of radioactive materials in the soil, drinking water, or building materials from which houses are constructed.

The people who live in these special areas of the world are obviously of considerable interest because they and their ancestors have been exposed to abnormally high radiation levels over many generations. If a radiation exposure of a few millisieverts per year is detrimental to health, causing genetic abnormalities or an increased risk of cancer, it should be evident in these people.

There are five major inhabited areas where there is a considerably increased amount of radiation from soil or rock; they are Brazil, France, India, Niue Island and Egypt.

In Brazil, several areas of the country, principally coastal strips each several kilometers long and several hundred meters wide, experience dose-rates from the soil or rock of 5 millisieverts (500 millirems) per year. Approximately 30,000 people are exposed to this level of irradiation and have been in perpetuity.

About one-sixth of the French population, that is, 7 million people, live in areas where the rocks are principally granite, which give rise to background radiation amounting to 1.8 to 3.5 millisievert (180 to 350 millirems) per year.

In Kerala and the Madras States of India, a coastal region 200 kilometers long and several hundred meters wide lies above an area of intense radioactivity, such that a population of over 100,000 people receive an annual dose-rate which averages 13 millisieverts (1,300 millirems) per year. This is the highest level of natural background radiation to which any human beings are exposed.

On Niue Island in the Pacific, a combination of volcanic soil and an unusually high radioactive content of plants result in a few thousand people being exposed to an external dose rate of 10 millisieverts (1,000 millirems) per year. In addition, these people ingest large amounts of radioactivity from plant material.

In the region of the Northern Nile Delta, a densely populated area of Egypt, dose-rates of 3 to 4 millisieverts (300 to 400 millirems) per year have been noted in several of the villages.

These high radiation levels should be assessed in the context of the *average* background to which most of the people of the earth are exposed, namely, 0.95 millisieverts (95 millirems) per year. This is the sum of contributions from cosmic rays, radiation from soil, rocks and the building material of houses, as well as from ingested radioactive isotopes. It is very much an average quantity.

The people who live in these areas of the world where the natural radiation is much higher than average have been carefully studied to see whether genetic anomalies, or cancer, are more prevalent than usual. So far, it has not been possible to establish any connection between the level of background radiation and an increase in biological disorders. It would be foolish, however, to derive too much comfort from this negative result, since the few studies that have been made were beset with difficulties. To begin with, the number of people who live in these areas of high natural background is relatively small. A more serious difficulty is that, as a rule, the people exposed differ from the majority of people on earth in such basic things as diet and social habits as well as ethnic origins. They also tend to live in small closed communities, and suffer from the increased congenital anomalies associated with inbreeding.

The kilt-wearing, bagpipe-playing Scotsman, living in the cold harsh climate of the Highlands, has little in common with the Indians of Kerala, who live in tropical fishing communities on the beach. Both are oddities compared with the average city-dweller of London, New York or Moscow. If a careful survey were to reveal small differences in the incidence of genetic anomalies and/or cancer, it would be naive to ascribe them to the small difference in background radiation while ignoring all of the other substantial variations.

Nevertheless, in spite of the obvious difficulties involved, it is a fact that no elevated levels of genetic anomalies or cancer incidence can be linked with high background radiation levels, even though some people live in areas receiving 10 times the average radiation doses. This has an important practical consequence. It is justification for believing that man-made radiation, in amounts comparable to background, is unlikely to produce a detectable number of detrimental biological changes in the world's population. This is a compelling argument.

If by living in the United States and receiving man-made radiation from nuclear power reactors and diagnostic x-rays, one does not receive a dose which is more than that received naturally by millions of people in France or hundreds of thousands of people in India, then it is difficult to imagine that any disastrous biological consequence will result.

To bring the story home: It is possible to live in New York City and have several x-ray pictures per year, without accumulating a total dose greater than that received naturally by everyone who lives in Denver, Colorado, where the cosmic ray background is unusually high. Since the people in Denver do not appear to suffer any ill effects from the elevated exposure levels in that place, there is some justification for assuming that a few medical x-rays taken in New York City will not be harmful.

5

The Not-So-Friendly Skies

THE RADIATION ENVIRONMENT ON EARTH

In common with most forms of life, human beings can only survive in the relatively temperate and hospitable conditions which exist on the surface of the earth. Unprotected, they would not last for one moment if subjected to the rigors of outer space, the heat by day, the cold by night, the lack of oxygen and moisture, and lastly the radiation levels. The surface of the earth constitutes a very special environment in which man has evolved for countless ages, and of which he is now a prisoner. If he aspires to leave, however briefly, he must take this environment with him. The particular interest of this book is radiation, and the surface of the earth is special and unusual in this, as in every other, respect.

Throughout most of the universe, radiation levels are much higher than on the surface of the earth, and there is an abundance of charged particles which would be damaging, if not lethal, to man. It is a hostile environment. The modest radiation levels on earth are due to two factors:

First, the earth behaves like a giant magnet, so that some charged particles are deflected and miss the earth altogether; others are deflected away from equatorial regions and funneled into the polar regions, and spiral down the lines of magnetic force, directly over the geomagnetic pole, where few people live. (See Figure 4.2 in chapter 4.)

Second, a thick atmosphere of air and water vapor envelops the earth like a blanket, breaking up, slowing down or stopping many of the fast particles coming from space.

But for these factors, the level of radiation on the surface of the earth would be much higher than it is. As soon as man engages in "unnatural" pursuits, such as space exploration, or flying high above the earth in a jetliner, he leaves the sheltered safety of the earth's surface and is exposed to some of the hazards of inhospitable space. This source of radiation is not, of course, man-made. It is perfectly natural and has always existed. However, it is "artificial" in the sense that man is exposed to it only as a result of his technological inventions. In the "natural" course of events he would stay on the surface of the earth where he belongs and where he is shielded.

It is first necessary to review the nature of the space radiations in order to understand their potential harm to the space traveler, or to the airline passenger.

SOURCES OF COSMIC RADIATION

The natural background radiation received by humans on the surface of the earth is made up of three components: gamma radiation from radioactive minerals in the earth, radiation from traces of radioactive elements in the body tissues resulting from the food we eat, and cosmic radiation. A representative mean value of the total dose equivalent from the sum of these natural sources is about 0.95 millisievert per year (95 millirem per year).

A space crew ascending from the launch pad through the earth's atmosphere is quickly free of the low level of dosage resulting from radioactive minerals in the rocks in the earth, but instead has now become subject to the much higher dose levels of cosmic radiation because they are no longer protected by the earth's atmosphere and magnetic field. Cosmic radiation comes from three different sources. First, there is the galactic radiation, which comes from far away in space, perhaps outside our galaxy, certainly from outside our solar system. Second, there is the radiation due to charged particles that are trapped in layers circling the earth. Third, there are occasional bursts of radiation coming from the sun in what are called solar particle events. These will be discussed in turn.

GALACTIC RADIATION

The galactic radiation has been studied for many years and is very familiar to physicists, but its origins are something of a mystery, because it comes from far out in the deep reaches of space. Especially puzzling are the tremendous energies of some of the particles involved, which cannot be equaled by even the most powerful man-made machines on earth. This radiation consists mostly of protons and alpha particles, i.e., atoms of hydrogen and helium from which the planetary electrons have been stripped, moving at incredible speeds close to that of light. Nuclei of most of the heavier elements are also present but in much smaller quantities. For example, approximately one particle in 2,000 has a mass greater than 20 times that of hydrogen; among these heavy particles, iron nuclei are quite abundant.

The dose-rate due to the galactic radiation does not vary much with time, and is not high enough to be lethal to astronauts, or to cause concern on the basis of our long experience with x-rays. However, the high-energy heavy nuclei are a form of radiation that is unique in space, and until recently could not be produced and studied on earth in man-made machines. The experience gained with x-rays may not be relevant to these heavy particles.

During the long trip to the moon on the Apollo 11 mission, one of the astronauts reported "seeing" light flashes while resting in a completely darkened cabin. Once alerted to the possibility, almost all of the crew members on subsequent lunar missions saw similar light flashes at the rate of one or two per minute, which they variously described as streaks, stars or diffuse luminous clouds. All of this was seen while their eyes were closed, and they were in complete darkness. This is due to high-energy particles passing through the eye, causing a direct effect on the retina, and has subsequently been simulated in laboratory experiments with fast neutrons and charged particles produced in an accelerator. These particles have such enormous energy that they readily penetrate the walls of a spacecraft and pass right through the human body. There is no possible way to shield astronauts from these particles.

A special category of the particles encountered in space have a very high energy and are large and heavy, i.e., they have a high atomic number (Z); these are known as HZE particles. On a 14-day round trip to the moon, it has been estimated that about 100 of these HZE particles traverse the head of an astronaut. Each particle destroys a number of the functional brain cells located along its path. The cause for concern is that the basic nerve cell, the neuron, does not divide in the adult and therefore losses are not replaced. In ordinary life, there is a continual small loss of cells in the brain as a function of age. The number of brain cells killed by the passage of HZE particles during space flight may be no greater than the number that die naturally during the course of life. What is special and unusual is that all the cells killed by an HZE particle are contiguous. The overall concept is of a column of dead or heavily irradiated cells along the track of the particle, surrounded by a cylindrical shell of injured or lightly irradiated cells. In simple terms, the effect is rather like having a number of very fine needles pushed through the brain. While this has produced no deleterious effects which are obvious during 14-day space missions, there is some concern that longer space journeys of several years duration would result in impairment of mental capabilities, or other malfunctioning of the nervous system.

These high-energy heavy particles are no problem to us on earth. Most are deflected by the earth's magnetic field, while the remainder have to pass through the atmosphere where they are broken up and attenuated by multiple collisions with atoms and molecules of the atmosphere.

RADIATION BELTS

Around the earth are regions in which the earth's magnetic field has trapped vast numbers of charged particles, which are forced to go back and forth from pole to pole and around and around in closed trajectories. These radiation belts were predicted long before actual observations were possible, although it was not foreseen that they would be so extensive, or contribute such large doses. (See Figure 4.2 in chapter 4.)

In going vertically up from sea level at the equator, the dose-rate rises rapidly at an altitude of about 700 miles, and persists to an altitude of 7,000 miles or more. Fortunately, much of this radiation consists of electrons and protons which do not have very high energy, and are stopped by the aluminum walls of a spacecraft. Within the radiation belt, the dose to an astronaut, inside of a spacecraft, could be of the order of 0.1 Gray/hr (10 rads/hour) or more. An exposure of several days at this dose-rate would

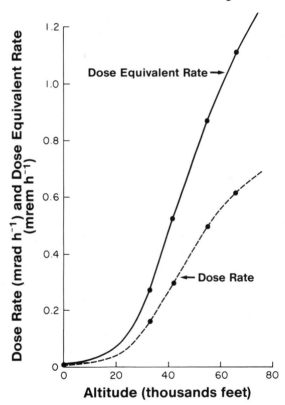

FIGURE 5.1. Amount of galactic cosmic radiation (dose-rate and dose equivalent rate) as a function of altitude at a latitude of 70°N. (Redrawn from Cosmic Radiation Exposure in Supersonic and Subsonic Flight. Final report of the Advisory Committee for radiation Biology Aspects of the SST. Aviation Space and Environmental Medicine, September 1975).

FIGURE 5.2(a). The space shuttle COLUMBIA, together with its liquid fuel tank and two solid rocket boosters, clears the launch tower at 10:59 a.m. on June 27, 1982. Astronauts Thomas C. Mattingly II and Henry W. Hartsfield are aboard. Minutes later, Columbia established an orbit at a height of 241 kilometers (130 nautical miles) above the surface of the earth, a height chosen to avoid the intense radiation from charged particles trapped in the Van Allen belts. Each crewman carried a personal dosimeter while other dose-measuring devices were located at various positions throughout the shuttle.

FIGURE 5.2(b). Touchdown! Columbia lands at Edwards Air Force Base, California, on Independence Day 1982 after a week in space. (Courtesy of the National Aeronautics and Space Administration (NASA) of the United States of America).

prove to be lethal, and therefore the time spent in this region must be severely limited.

Space missions which involve orbiting the earth for weeks or months, such as the U.S. space shuttle, do so at lower altitudes of a few hundred miles above the earth to avoid the problem posed by these trapped radiations. Even then, missions are limited in duration by the South Atlantic anomaly. As its name implies this is a region over the South Atlantic where the radiation belts dip much more closely to earth than at any other place. This is due to a lack of symmetry in the magnetic field of the earth. Any mission comprising a large number of revolutions around the earth must pass through this anomaly during some orbits, and since the dose-rate is much higher in this anomaly than elsewhere, the total radiation dose gradually accumulates to objectionable levels in a few months, unless steps are taken to provide extra shielding. As it turns out, weightlessness produces physical and metabolic problems too, including a reduction in heart size and weakened muscles, so that the duration of flights must be limited to about four months for these reasons. The position is much worse for the Russians than for the Americans, since, lacking the convenience of the space shuttle, they must expect their crews to remain in space for protracted periods whereas the Americans can change personnel regularly and minimize the deleterious effects of space.

The situation is quite different for a deep space venture, such as a trip to the moon, in which case it is not possible to avoid passing through the radiation belts. However, a spacecraft launched from earth with escape velocity is going so fast as it passes through the radiation belts that the time of exposure to the high dose-rate is limited to a few minutes.

SOLAR PARTICLE EVENTS

By far the greatest radiation hazard in space travel is the possibility of a major solar particle event. Ordinarily the surface of the sun is at a temperature of about 5,700 degrees, but occasionally a limited area of the surface reaches a million degrees. The activity begins as sun spots, and occasionally these turn into flares.

Flares develop rapidly in a matter of minutes and are impossible to predict except that they tend to run in 11-year cycles. Peaks in solar flare activity occurred in 1948, 1959, and 1970, while activity was at a minimum half-way through each of these 11-year solar cycles. These cycles merely represent trends and at no time can it be guaranteed that a solar event will not take place. Unlike galactic radiation, then, solar cosmic radiation is exceedingly erratic, although it is roughly related to the 11-year cycle of solar activity, characterized for years by sun spot activity as described above.

Cosmic rays from the sun are principally protons covering a wide energy range, although alpha particles may sometimes be present in significant amounts. The first charged particles may begin to arrive on earth about 15 minutes after a solar flare becomes visible on the surface of the sun, so that the warning time is short.

Solar events are very variable, but it is usually concluded that there may be one of two events of distinct interest each 11-year solar cycle. Bearing in mind that each event lasts several hours, the probable duration of significantly elevated dose-rates might be about 10 to 20 hours each 11 years. A major event occurred in August 1972, which, it is estimated, would have resulted in a peak dose equivalent rate at 20 Km (65,000 ft.) of about 3.5 millisievert/hr (350 millirem/hr.). The previous 11-year period included the great flare of February 23, 1956, which has been estimated to have produced a dose equivalent rate at an altitude of 20 Km of the order of a few rem/hr.

The dose-rate to an astronaut, of course, can be reduced by extra shielding material in the walls of the spacecraft, but considerations of weight limit what can be done in this regard. As a result, long-term space missions to date, such as those involving a trip to the moon, were chosen for periods when the solar activity was at a minimum. The possibility of encountering large and significant radiation doses in space is quite real, and has been a serious concern for manned space flight from its earliest planning stages. This will always be important, and will involve a considerable hazard and a great deal of difficulty in scheduling longer-term space missions in the future.

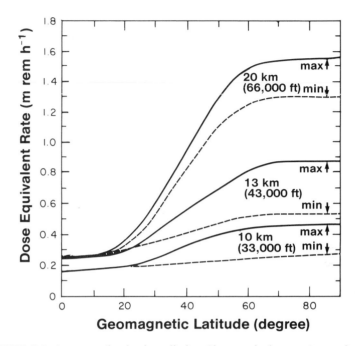

FIGURE 5.3. Amount of galactic radiation (dose equivalent rate) as a function of latitude and altitude for the maximum and minimum during the solar cycle. (Redrawn from "Cosmic Radiation Exposure in Supersonic and Subsonic Flight." Final report of the Advisory Committee for Radiation Biology Aspects of the SST. Aviation Space and Environmental Medicine, September, 1975).

The risk of serious harm from radiation was appreciated in the early U.S. space missions, particularly the chance of a sudden solar event, but it seemed at the time to be one of the lesser hazards compared with being launched from earth on top of a rocket, or being stranded on the moon hundreds of thousands of miles from home. In retrospect, however, due to the incredible safety record of the manned space-flight program, in which not one single life has been lost on a mission, the radiation exposure has turned out to be the only potentially harmful legacy which astronauts retain after the trip.

Individual crew members have received appreciable radiation doses, though not as much as some had predicted. It is unlikely that the number of space travelers will be large enough for the radiation they receive to add an appreciable genetic burden to the human race, especially if the practice is continued of choosing mature individuals for astronauts who have already produced their children. Any foray into space inevitably involves a small chance of a disaster from a massive dose of radiation resulting from a large and unexpected solar event.

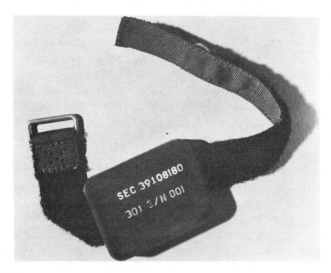

FIGURE 5.4(a). Dosimeter worn by the astronaut crew of Skylab to monitor radiation. (Photograph courtesy of the United States National Aeronautics and Space Administration.)

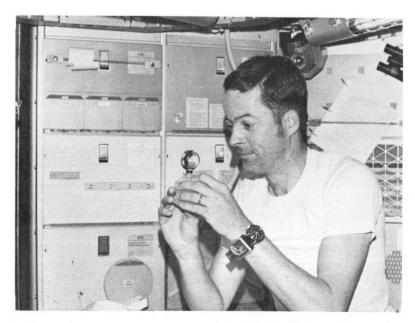

FIGURE 5.4(b). Dr. Joseph Kerwin working in Skylab while orbiting the earth; note the dosimeter on his left wrist. (Courtesy of the United States National Aeronautics and Space Administration.)

HIGH-FLYING JETLINERS

A transatlantic flight from the United States to Europe involves an extra exposure to cosmic radiation amounting to about 0.05 millisievert (5 millirems). By climbing to an altitude of around 35,000 feet, the shielding effect of part of the earth's atmosphere is lost, and as a result passengers and crew receive this small additional dose of radiation. For someone who completes, say, four round trips across the Atlantic per year, this amounts to a total of 0.4 millisievert (40 millirems). This is approximately equal to the difference in natural background radiation received by a resident of Colorado compared with a New Yorker. The pilot and crew of jetliners, making frequent flights across the Atlantic or across the continent, may regularly accumulate more than the 5 millisieverts (500 millirem) per year, which is the maximum permitted for members of the general public. Technically they should be designated as radiation workers, though in practice they never are. The doses they receive are, on average, higher than those incurred by the vast majority of workers in most other occupations, including hospitals or the nuclear power industry, though their radiation exposure is unlikely to reach the higher levels encountered by a few specialist individuals as, for example, technicians refueling a reactor or operators in a fuel reprocessing plant. The radiation to which they are exposed is not "man-made " because of course it is natural in

FIGURE 5.5. The Anglo-French supersonic jetliner Concorde in flight. (Courtesy of the British Aircraft Corporation.)

origin and has always been there. The risk is man-made in as much as human beings would not, in natural circumstances, spend their working life 35,000 feet above the surface of the earth.

SUPERSONIC AIRCRAFT

There will be additional factors to be considered concerning the radiation exposure of airline passengers now that supersonic aircraft, such as the Anglo-French Concorde, have come into routine service. In order to operate efficiently, Concorde must fly at 50–60,000 feet, where the level of cosmic radiation is appreciably higher than at the altitudes used by the present generation of jumbo jets. The United States SST, before it was cancelled, was

FIGURE 5.6. Photograph of the radiation-measuring device fitted on the Anglo-French supersonic jetliner Concorde. (Courtesy of Mr. H.A. Davenall, British Aircraft Corporation Ltd.)

FIGURE 5.7. Control panel of the third crew member of Concorde, showing the radiation monitor, in the bottom right-hand corner. (Courtesy of the British Aircraft Corporation.)

scheduled to fly even higher at 80,000 feet. Under normal circumstances, however, the maximum dose received by a passenger on a supersonic transatlantic flight would be no more than a conventional jetliner (i.e., about 50 microsievert or 5 millirem) because, although the dose-rate is higher, the duration of the flight is halved. The accompanying figures show how radiation dose rate varies with altitude and latitude (see Figs. 5.1 and 5.3).

A new factor introduced by the higher altitude is that solar events can no longer be ignored. If an exceptionally large event occurs, a significant radiation dose could be received by the passengers and crew in such a plane. For this reason, the Concorde carries dose-measuring instruments to alert the pilot if the radiation reaches an unsafe level. Should the dose-rate reach 0.5 millisievert/hr (50 millirems/hr), the supersonic jetliner would descend to a lower altitude to enjoy the screening effect of several miles of atmosphere and continue its journey at subsonic speed. Large solar events, producing radiation doses sufficient to warrant such actions, would only be expected to occur a few times during each 11-year solar flare cycle. However, if the day ever comes when a large number of supersonic aircraft are in routine service, then it is sure that a few of them would be in the air during a solar event, and it is prudent to make provision to protect the traveler in such a contingency.

Longterm, it is clearly desirable that a system be developed in which radiation data obtained from satellites should be monitored and used to warn of any unusually high exposures to the crew and passengers of high-flying aircraft.

6
The Healing Rays

RADIATION IN MEDICINE

It is not generally appreciated that there are three quite separate applications of radiation in medicine that exploit different properties of radiation and have divergent objectives from a medical point of view. They may be summarized as follows.

1. Radiation is used to diagnose disease. The dentist may take x-ray "pictures" to seek hidden cavities in teeth; the doctor may use x-ray pictures to locate a broken bone or to search for some more subtle abnormality in the body. This is called *x-ray diagnosis*.
2. Radioactive isotopes are administered to some patients. These isotopes give off radiation and may be detected outside the body after they have been localized in certain specific organs deep within the body. The purpose of such tests may be to assess whether body functions are normal or abnormal, or to detect the presence of a tumor. This use of radioactive materials for diagnosis or treatment is called *nuclear medicine*.
3. Radiation may be used, not to detect disease, but to treat it. These days, radiation as a form of treatment is reserved almost exclusively for malignant diseases, i.e., cancer. The purpose of the radiation is to kill the malignant cells. This is called *radiation therapy* or *radiation oncology*. Oncology means the study of tumors.

X-RAY DIAGNOSIS

X-rays were discovered by the German physicist Wilhelm Conrad Roentgen in the year 1895. Roentgen was a physicist experimenting with gas discharge tubes. He discovered that the combination of a low pressure of gas inside the tube, and a high voltage across the tube, produced mysterious rays which were emitted from the cathode (the negative terminal). These rays could blacken photographic film and had the ability to pass through materials opaque to light, such as paper, wood and even thin sheets of metal. He did not know what they were, so his first publication describing them was entitled "On a New Kind of Ray"; they were later called x-rays.

Radiation and Life

FIGURE 6.1. Wilhelm Conrad Roentgen, the discoverer of x-rays, as Rector of Wurtzburg University. (Courtesy of Eastman Kodak)

Roentgen was a physicist and had no connection with hospitals or medicine. However, when giving a lecture to demonstrate x-rays, he asked von Kulliker, who introduced the lecture, to put his hand in front of the x-ray machine, and with a sheet of photographic film made the first x-ray picture or radiograph (Figure 6.2), which shows the bony structure of his hand, as well as his ring!

The enormous potential of x-rays in medicine was appreciated within a matter of months, and the techniques developed very rapidly, first in Germany and then in all of the countries of the Western world. Before the close of the century, a mobile x-ray unit had even accompanied the British army to the Sudan to help locate bullets and shrapnel in soldiers wounded at the battle of Omdurman (Figure 6.3). X-rays are now one of the most powerful tools available to the doctor for diagnosing disease, and one of the most widely used. In the United States, about half of the population is x-rayed every year for medical purposes. In the last comprehensive survey carried out in 1970, 76

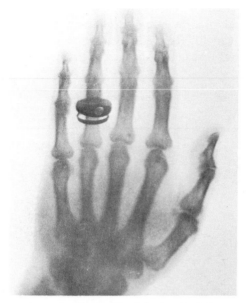

FIGURE 6.2. X-ray picture (radiograph) of the hand of the anatomist von Kulliker taken by Roentgen in January 1896—just a few months after the discovery of x-rays. This is the first radiograph of a living object and opened up the whole field of "radiology"—the medical use of x-rays. (Courtesy, Roentgen Museum, Wurtzburg, Germany.)

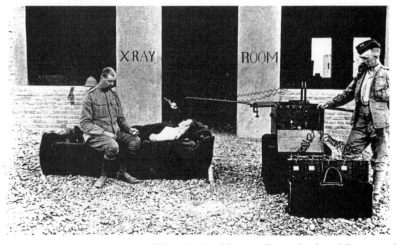

FIGURE 6.3. Major Battersby and his orderly taking a radiograph of a soldier wounded in the battle of Omdurman in the Sudan, September 2nd, 1898. The first military use of x-rays coincided with the last cavalry charge of the British Army, in which, incidentally, Sir Winston Churchill participated. (Courtesy of Dr. H.H. Saxton and the Editors of the British Journal of Radiology.)

million persons had an x-ray examination, and 59 million in addition had
dental x-ray examinations. The pattern varies somewhat from one country to
another but is similar throughout the industrialized world.

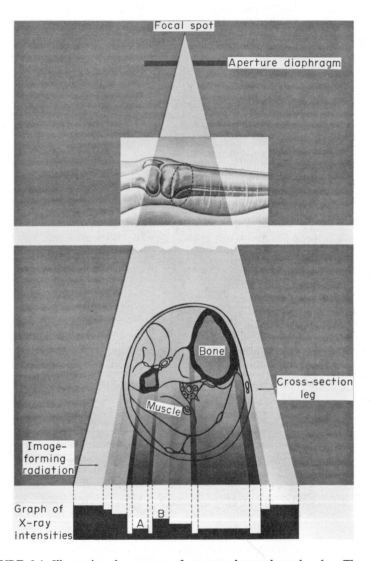

FIGURE 6.4. Illustrating the passage of an x-ray beam through a leg. The x-rays
produced at the focal spot penetrate the tissues in different amounts. Center. A cross-
section of the leg indicates in detail how the structures within the leg cause a variation in
the intensity of the image-forming radiation. Below. A graph of the x-ray intensities as
they fall on the film. (Courtesy of Eastman Kodak.)

The Principle of an X-Ray Picture

The principle of a radiograph is based upon the ability of x-rays to pass through the human body, whereas visible light cannot; x-rays therefore enable the observer to "see" within the body. The principle of an x-ray picture is illustrated in Figure 6.4. As the rays pass through the body, they are stopped to a greater extent by materials which are dense or which are composed of elements which have a higher atomic number.* In general, x-rays pass through soft tissue most easily and are stopped most by bone. Not only is the bone more dense than soft tissue, but it also contains a larger proportion of calcium, which has a higher atomic number. The film, placed behind the

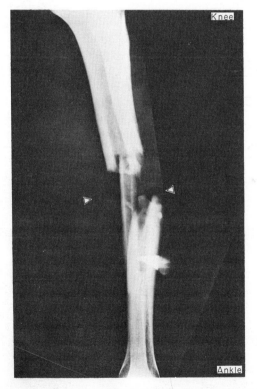

FIGURE 6.5. A broken bone in the leg between the knee and ankle. (Courtesy of Dr. W.B. Seaman, Columbia-Presbyterian Medical Center, New York.)

---

*The atomic number (z) corresponds to the position of the element in the periodic table. Hydrogen is 1, while uranium is 92. The atomic number is equal to the number of protons in the nucleus and therefore to the number of electrons in orbit around the nucleus.

body, is blackened to an extent dependent on the amount of x-rays it receives. It will be blackest where the body is thin and composed of soft tissues. Bones show up as light shadows because they absorb some of the x-rays and prevent them from reaching the film. To the untrained eye of the layman, the bony structures are the only things which show up clearly on a radiograph; but to the experienced eye of the radiologist, who has spent half of his life interpreting the shadows on an x-ray picture, more subtle details of anatomical structure become evident.

Until recently the majority of x-ray images were produced on film. Because of the number of films used and the size of the films (a typical chest radiograph, for example, measures 14 × 12 inches) a large proportion of the world supply of silver was used for the production of the film emulsion. This is now changing. In many applications the x-rays fall instead on an array of sensitive crystals which convert the image contained in the pattern of x-rays into electrical impulses. The resultant picture can then be displayed on a television screen and processed by electronic devices to alter the contrast or brightness of the picture in order to extract more detail and information from it. This is called *digital radiography.* Another advantage of this innovation is that instead of storing a large packet of heavy films for each of the thousands of patients that each year pass through a major medical center, the same amount of information can be stored on a few video discs. Patient x-ray

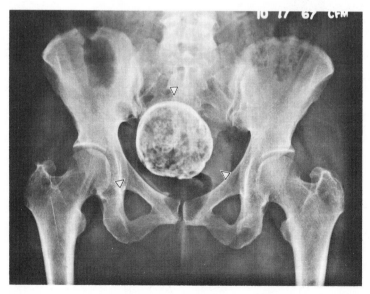

FIGURE 6.6. A huge calcified tumor of the uterus is clearly seen in this picture. (Courtesy of Dr. W.B. Seaman, Columbia-Presbyterian Medical Center, New York.)

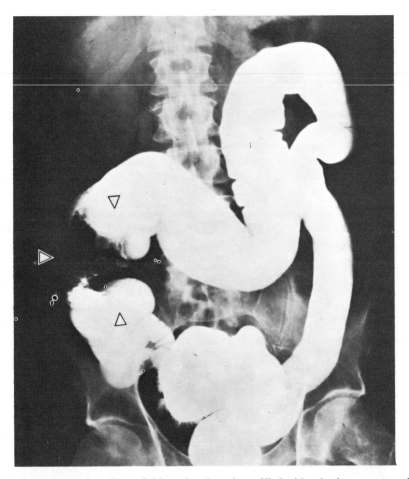

FIGURE 6.7. The intestines of this patient have been filled with a barium compound which is opaque to x-rays. The blockage caused by a cancer is then clearly visible. (Courtesy of Dr. W.B. Seaman, Columbia-Presbyterian Medical Center, New York.)

information in this form can also be transmitted over a telephone line from one hospital to another—across town or across the country. The computer age is revolutionizing radiology.

Seeing the Invisible

A few examples of the extreme usefulness of x-rays in diagnosing disease, or detecting abnormalities, are illustrated in the accompanying pictures. A broken or cracked bone is at once obvious in an x-ray, even when it may be difficult to diagnose by direct examination due to the presence of swelling and extreme tenderness. Figure 6.5 shows an x-ray of a fractured bone in the leg.

Locating the exact position of a foreign object, such as a bullet or piece of shrapnel, is also a relatively simple matter and an invaluable guide to the surgeon, whose job it is to remove it.

X-ray pictures are particularly useful to diagnose abnormalities deep within the body, where direct physical examination would not be possible except by extensive surgery. An important factor is that a number of x-ray studies may be performed on a sick patient in order to arrive at a diagnosis, with no risk to his life and a minimum of discomfort and trauma. The presence of a "stone" in the kidney or gall bladder, for example, may be

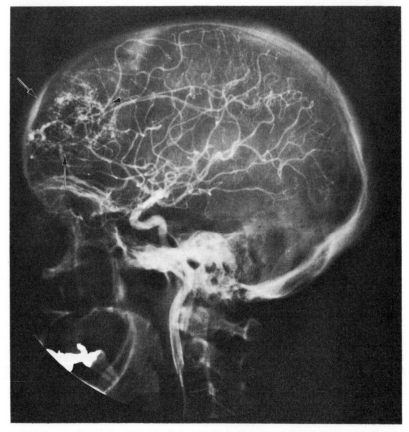

FIGURE 6.8. Examples of the use of contrast material, opaque to x-rays, in the brain. (a) In this patient, a material opaque to x-rays has been injected into the blood supply to the head. The distorted pattern of arteries indicates the presence of a brain tumor. (Courtesy of Dr. W.B. Seaman, Columbia-Presbyterian Medical Center, New York.)

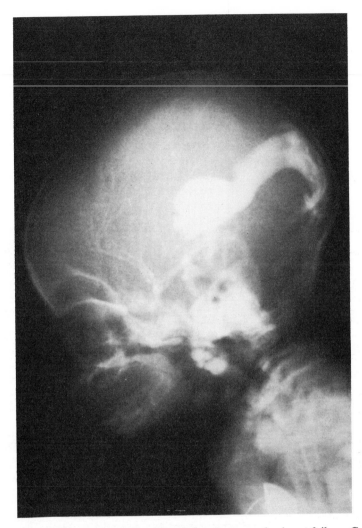

FIGURE 6.8. (b) Lateral skull film of an infant in congestive heart failure. Contrast material, injected into an artery leading to the head, is seen to pass directly from the arteries to the veins; this abnormality represents an extra-cardiac shunt—i.e., short circuits the heart—which is responsible for the heart failure. (Courtesy of Dr. David Baker, Professor and Chairman of Radiology, Columbia-Presbyterian Medical Center, New York, N.Y.)

demonstrated by a series of radiographs. A huge tumor in the uterus is illustrated in Figure 6.6. It is like a large ball, 4 inches in diameter, and is very clearly visible because it is calcified; since calcium has a relatively high atomic number, it shows up clearly on an x-ray picture.

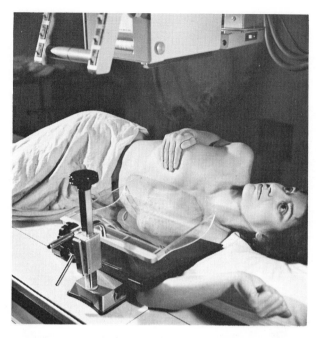

FIGURE 6.9. (a) Patient positioned for an x-ray examination of the breast. The x-ray machine is located above, and the black cassette under the breast contains the plate which records the x-ray image. (Pictures by courtesy of the Xerox Corporation, Pasadena, California, USA.)

Obstructions or abnormalities in the digestive tract show up clearly only when special techniques are used. Since the walls of the stomach and intestine are all composed of soft tissue, little detail can be seen on a straightforward radiograph. Instead, the patient is fed a barium mixture which is relatively opaque to x-rays. When the digestive tract is full of the mixture, its outline shows up clearly. An ulcer in the duodenum, or a blockage in the intestine due to a tumor, can frequently be demonstrated in this way; one such example is shown in Figure 6.7. Dynamic investigations can also be made, such as a study of the rate of emptying of the stomach. A number of pictures are taken, several minutes apart, as the opaque barium passes through and out of the stomach.

Radio-opaque materials can be injected into the blood stream to highlight irregularities in the pattern of veins and arteries. For example, the presence of a tumor in the brain can sometimes be proven, and its position located, by noting a distortion in the pattern of blood vessels made visible by a radio-opaque material. A case of this type is also illustrated in Figure 6.8. The lymph system can also be visualized by using a radio-opaque material, which will show up irregularities or obstruction.

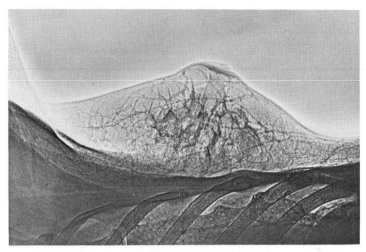

FIGURE 6.9. (b) Xeromammograph of a normal breast.

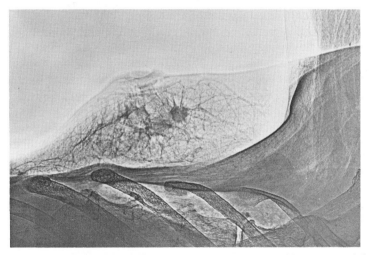

FIGURE 6.9. (c) Xeromammograph of a 35-year-old woman with a cancer of the left breast. Note the characteristic skin thickening and nipple retraction.

Mammography

A particular use of x-rays for diagnosis which has been much in the news in recent years is mammography—the technique whereby x-ray pictures of the breast are taken in order to detect the possible presence of cancer.

Mammography is the only screening technique for breast cancer that has proved to be successful so far, inasmuch as it does show up tumors of less than 5 mm in diameter that could not be detected by any other means. Since the breast is composed entirely of soft tissue with no bony structures, special low-voltage machines are used to show up small differences in the composition of muscle and fat, and to highlight small irregularities that could indicate a benign cyst or a malignant tumor. Tumors as small as 5 mm usually show up only if there is "micro calcification with branching"—i.e., tiny deposits of calcium which fan out like branches on a tree. Calcium shows up clearly on an x-ray picture because it has a higher atomic number (z) than soft tissue or fat. The typical appearance of a speculated mass is shown in the accompanying photograph (Figure 6.9c). The routine use of mammography to screen large numbers of women for the possible presence of cancer has been the subject of much debate over the years.

For mammography to be useful it must be carried out by a specialist, using equipment designed for the purpose. Low voltage x-ray machines, together with sensitive films and screens, have now been developed to the point where a mammographic examination can be carried out (2 films for each breast) with the radiation dose not exceeding 2 milligray (0.2 rad) This is 10 to 100 times less than it was a decade ago.

It is perhaps of interest to review the results of some of the large screening programs that have been carried out. In Great Britain, Marks and Spencer (a large chain store) arranged for 21,000 of their female employees to be screened, which revealed 103 cases of previously unsuspected cancer. This corresponds to a detection rate of about 5 cases/1000 for all women over 35 years of age, and about 8 cases/1000 in the older age group of 50 years old and above. A much larger study in the United States involved 261,000 women. The detection rate was about 5/1000. On average, about one-quarter of the cancers detected were small—i.e., less than 1 cm in diameter, but this varied widely from one state to another, from a low of 13 percent to a high of 80 percent.

While there are differences of opinion among the "experts," it is clear from the above discussion that mammography is a valuable method for detecting breast cancer in its early stages. The current advice is for women with no symptoms to have a baseline mammographic examination between ages 35 to 40, repeated subsequently at one to three year intervals unless there are grounds for suspicion. After age 50, a mammograph should be included in annual checkups.

The CT Scanner

The practice of diagnostic radiology has been revolutionized since 1972, due to the introduction of Computer Assisted Tomography. The machine that makes this possible—the CT scanner—was initially developed to allow a

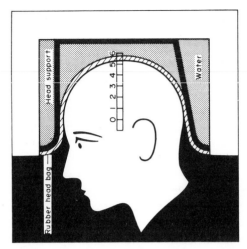

FIGURE 6.10. The patient positioned for examination with the EMI scanner. The head is held in a rubber head-bag inside a water-filled cube. (Courtesy of EMI Electronics and Industrial Operations.)

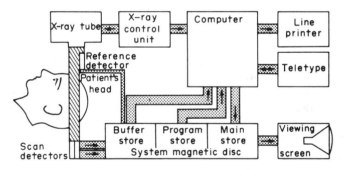

FIGURE 6.11. In the EMI scanner, the head is examined as a series of "slices." (Courtesy of EMI Electronics and Industrial Operations.)

better visualization of the brain. The x-ray machine rotates in a circle around the patient's head (see picture and figures) in the course of which many thousands of x-ray exposures are made with a narrow pencil beam of x-rays. Each time the beam of x-rays passes through the head, it is stopped to a greater or lesser extent depending on the amount of bone or tissues of varying density in its path. Instead of falling on a film, the x-rays activate a crystal detector which passes a signal to a small computer. The computer collects all of the information, reconstructs the distribution of materials of different density and atomic number within the skull, and displays the result on a television screen. In particular, sudden small changes of density corresponding

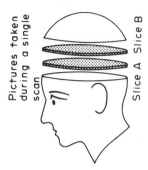

FIGURE 6.12. Block diagram of the basic EMI scanner system. (Courtesy of EMI Electronics and Industrial Operations.)

to boundaries between different structures show up clearly. The whole procedure takes a few minutes.

The result is a startlingly clear visualization of a "slice" of the skull. The detail that can be seen far exceeds anything that is possible with a conventional x-ray picture. This ingenious machine was first developed by a relatively small subsidiary of a British company (Electrical and Musical Instruments) and could be used only for the head. New and improved models have been developed by major companies around the world that can produce pictures of any part of the human body. Examples are shown to illustrate the detail with which internal organs can be visualized and any abnormalities identified by the practiced eye of the radiologist (see Figs. 6.14, 6.16, and 6.17). These machines are expensive (about a million dollars) and can only be justified in major medical centers where they can be fully utilized.

DEVELOPMENTS IN DIAGNOSTIC RADIOLOGY

Until the early 1970s, the basic techniques used in diagnostic radiology had not changed much in over half a century. X-rays generated by a high-voltage device passed through the patient and produced an image on a film. This is perhaps stating it too simply but is not far from the truth. In the last decade or so, all this has changed. The marriage of x-ray beam to the computer brought about the CT scanner—which has been described as the greatest discovery in imaging since Roentgen discovered x-rays! It has revolutionized radiology and represents high technology. The manufacture of the equipment is a multimillion dollar business; large companies spring up, flourish and go out of business in a relatively short time if they cannot keep up with the flood of new ideas. Engineers, physicists, and computer experts are very much in demand in the design, development, and operation of the machinery of modern diagnostic radiology.

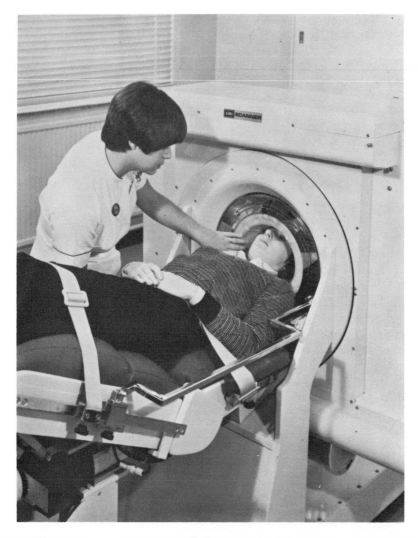

FIGURE 6.13. For examination by the EMI scanner, the patient relaxes on a couch with the head in a water-cushioned rubber head-bag in the center of the scanning unit. (Courtesy of EMI Electronics and Industrial Operations.)

Before most medical centers have finished paying for their latest CT scanner, a new revolution is gathering momentum—the introduction of imaging techniques based on NMR—Nuclear Magnetic Resonance. These devices involve no x-rays, but instead a very large magnetic field and radiofrequency waves. The basic principle is very simple. Certain atomic nuclei, such as hydrogen, behave as small magnets when placed in a uniform static magnetic

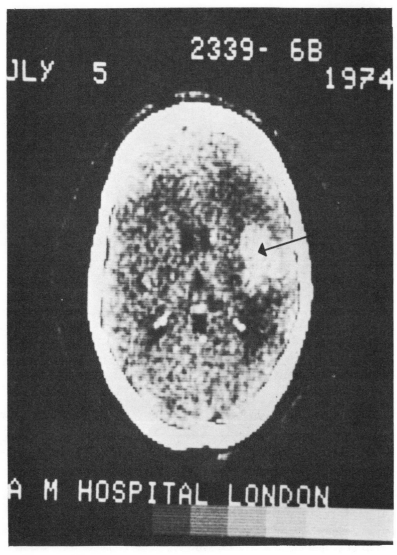

FIGURE 6.14. EMI scanner picture of a patient with a glioma (brain tumor) of the right temporal region of the head. (Courtesy of EMI Electronics and Industrial Operations.)

field; most will align themselves in the direction of the field. To induce nuclear resonance, a short radiofrequency (RF) pulse is applied via a coil surrounding the patient. The RF radiation is equivalent to the application of a second, much smaller magnetic field, which rotates around the static field. If the

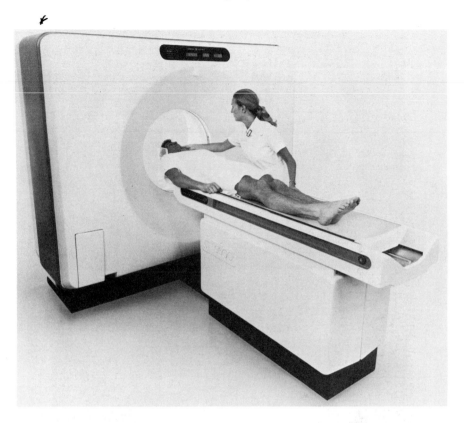

FIGURE 6.15. Photograph of a modern CT scanner by the General Electric Company that can take an x-ray picture of a section or "slice" through any part of the body. The photograph shows a technician positioning the patient; a section will be produced of that portion of the body within the circular ring. (Courtesy of General Electric Company.)

frequency is just right, this causes the atomic "magnets" to precess around the magnetic field—much like a spinning gyroscope precesses when its axis is tipped out of alignment with the earth's gravitational field. When the RF field is removed, the nuclei will line up with the static magnetic field again—i.e., return to the lower energy, more stable, state. The extra energy will be radiated and this can be detected. The time it takes for the return to the lower state is called the relaxation time, and depends on how readily the nuclei can dispose of their excess energy; this differs from one tissue to another. With conventional x-rays, the signal output is related directly to the fractional transmission of x-rays, and therefore reflects the density and atomic number of the tissue through which the x-rays have passed. In NMR imaging, the

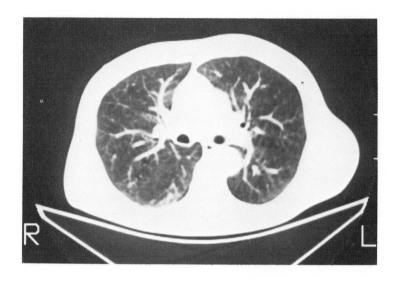

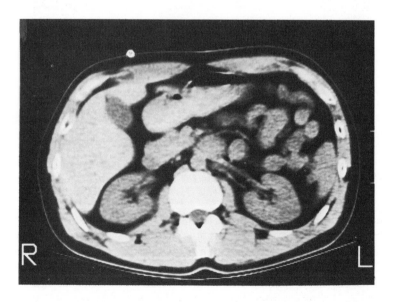

FIGURE 6.16. Pair of pictures produced by a General Electric whole-body scanner.
Upper Panel: Section through the chest showing good visualization of the lungs. Small vessels can be seen and emphasematous areas.
Lower Panel: Section through the pelvis showing the liver. (Courtesy of General Electric Company.)

signal depends on the concentration of resonating nuclei in the resonant volume or plane and on the relaxation times. As a consequence, nuclear magnetic resonance provides evaluation of the chemical and biological state of the tissue of interest. The information provided is different from that in a conventional x-ray and should provide greater understanding of disease. It has been demonstrated, for example, that neoplastic and other abnormal tissues exhibit longer relaxation times than the equivalent normal tissues. (For examples of NMR images, see Figs. 6.20 and 6.21.)

At the present time, many distinct imaging techniques have been proposed and demonstrated to process the NMR signals produced and form an image. It is not clear which will prove to be best.

Because of their abundance and strength of the signals, only protons have been studied in clinical NMR imaging trials to date. Field strengths up to about 0.2 Tesla are required, for which the resonant frequency is about 8.5 MHz. However, other biologically interesting nuclei that are NMR sensitive include carbon, flourine, sodium, potassium and phosphorous. Probably

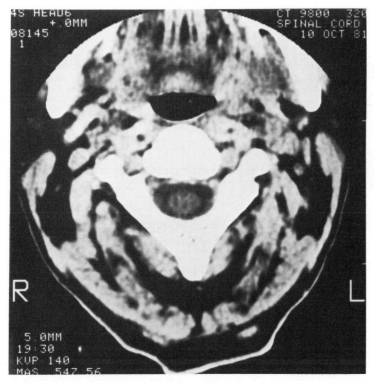

FIGURE 6.17. Section through the neck produced by a General Electric whole-body scanner. The spinal cord is clearly visible without the need for a contrast material. (Courtesy of General Electric Company.)

phosphorous will turn out to be most important because of its role in energy metabolism at the cellular level. To image phosphorous nuclei at 8.5 MHz requires a field strength of 0.5 Tesla, for which super-conducting magnets are required, i.e., magnets operating at very cold temperatures close to the absolute zero. The potential uses of NMR seem to be limitless.

The diagnosis of the presence and extent of cancer is an area actively under investigation at present. Proton determinations by NMR may be used to evaluate diseases related to water content and movement. Noninvasive determinations of blood flow have important clinical implications in patients with cerebrovascular disease. Another potential use of NMR imaging is in clinical evaluation of the energy cycle of cells from the concentration and distribution of carbon and phosphorous. Since basic cellular energy metabolism is related to particular diseases, NMR may one day prove to be a useful and effective screening technique.

Fabricating instruments to produce a static magnetic field with sufficient dimensions to allow insertion of the human torso has proved to be a formidable challenge. Two approaches have been tried; first, a permanent magnet. To

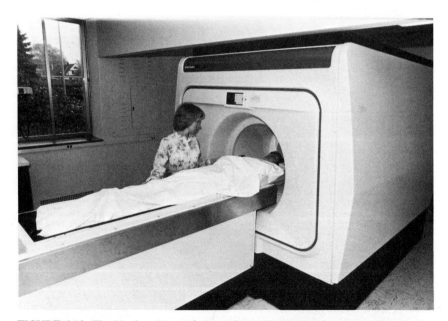

FIGURE 6.18. The Nuclear Magnetic Resonance (NMR) imaging device, installed in the Radiology Department of University Hospitals of Cleveland/Case Western Reserve University. The heart of the machine is a 1.5 kilogauss superconducting magnet (i.e. operated at the temperature of liquid helium). (Courtesy of Dr. Ralph Alfidi, Department of Radiology, University Hospitals of Cleveland/Case Western Reserve University.)

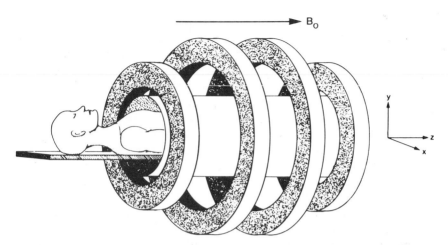

FIGURE 6.19. The inside lay-out of a four coil electromagnet used in most clinical NMR imaging applications. The portion of the patient's body to be imaged is positioned within the electromagnetic coil, which produces a high magnetic field. The size and weight of the electromagnet is minimized by operating it very cold—at the temperature of liquid helium.

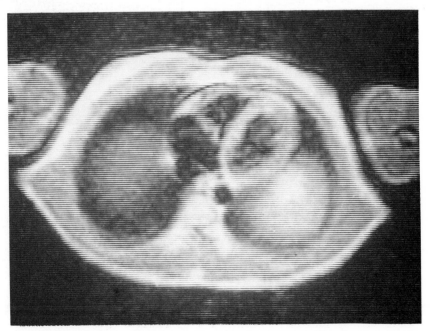

FIGURE 6.20. NMR scan through the chest, showing the details of the heart—the right and left ventricles—as well as the aorta. (Courtesy of Dr. Ralph Alfidi, Department of Radiology, University Hospitals of Cleveland/Case Western Reserve University.)

produce a field of a .2 Tesla in a cavity big enough for a human being requires a magnet weighing 100 tons—too much for the weight-bearing characteristics of a floor in a normal hospital. Also, a permanent magnet cannot, by its nature, ever be switched off! Operators and patients must be very careful to remove all metal objects before coming near a magnet of this sort. A screwdriver in such a powerful magnetic field becomes a lethal missile, and it is said that such a field could tear a pacemaker out of a patient's chest! The alternative is to use a cryogenic magnet, i.e., to produce the magnetic field by an electrical current, which becomes feasible only at very low temperatures— that of liquid helium. The device is not as heavy and can be turned off—but of course has a voracious appetite for expensive liquid helium and liquid nitrogen. A device of this kind is illustrated in Figure 6.18. When all of the technical problems are overcome, the result is the most remarkable pictures that threaten to make the CT scanner obsolete. Some examples are shown in the accompanying illustrations. There is the further promise that an even higher magnetic field would allow the imaging of other elements in the body.

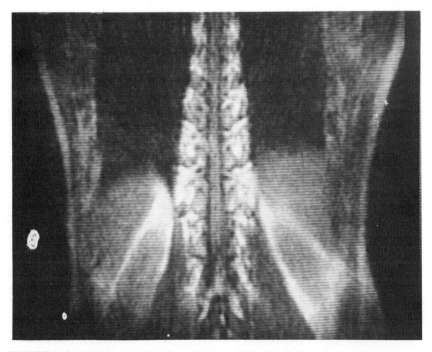

FIGURE 6.21. NMR scan showing with incredible detail the spinal cord and nerve roots. (Courtesy of Dr. Ralph Alfidi, Department of Radiology, University Hospitals of Cleveland/Case Western Reserve University.)

Experimental prototypes have been made with magnetic fields up to 2.5 Tesla. A note is in order concerning the biological hazards. The proponents of NMR list as one of the advantages that no x-rays are involved. This is true—but the known is replaced by the unknown. Instead of a small dose of x-rays, there is a high magnetic field and a pulse of RF energy. This could conceivably have biological implications. At levels proposed for clinical use, no adverse effects have yet been seen in the few humans studied or in experimental animals. However, as with any new modality, time will tell. Additional studies to get an early warning of biological effects are clearly required.

As a footnote to the development of these new diagnostic modalities, it is worth noting that as medical technology keeps improving and advancing, bringing a steady torrent of ever more sophisticated and expensive equipment (not necessarily more effective) and practices to market, the already high cost of medical care has been rising faster than ever. Nearly 10 percent of the gross national product in the United States goes to medical care. More than 85 percent of Americans are covered by medical insurance of some sort, so that doctors are encouraged to provide the most advanced—and expensive—tests and treatments available. All the incentives are to provide very high quality service at a very high cost! This must be a cause of considerable concern.

NUCLEAR MEDICINE

The availability of radioactive tracers has opened up a whole new field known as nuclear medicine, that has become a speciality in its own right. Nuclear medicine is based on the use of *radioactive isotopes*. Atoms are isotopes if they have the same number of protons in their nuclei, but different numbers of neutrons. They are variants of the normal element. Because they have the same number of protons, they also have the same pattern of electrons in orbit, and consequently their chemical properties are identical. As a result of having too many or too few neutrons, some isotopes are radioactive, which means that they give off radiation which can be detected with a sensitive instrument such as a Geiger counter or a scintillation counter. The radiation may be gamma rays or beta rays, or both. These are the isotopes used in nuclear medicine.

When a radioactive isotope is given to a human, it is possible, by measuring outside of the body with a counter, to detect where the isotope is localized, in what quantity, and the pattern of distribution. This information is invaluable in the diagnosis of a number of medical problems.

Very small amounts of radioactive materials can be given in nuclear medicine because of the great sensitivity with which radiation can be detected. As a result, tests can be performed with satisfactorily low radiation doses to the tissues, but it also means that very small *masses* of material need to be

FIGURE 6.22. The great Hungarian chemist, Hevesy, whose work beginning before the First World War earned him a Nobel prize in 1943. He was the first to conceive of using radioactive isotopes to label compounds for biology and medicine. (Courtesy of the University of California Lawrence Berkeley Laboratory.)

administered. In many body processes, especially those involving hormones or vitamins, the normal balance is easily disturbed. Radioactive tests seldom require more than a microgram (one millionth of a gram) of the substance under test, which will not disturb the normal balance. This is an important asset of radioactive isotopes for use in medical and biological studies.

Thyroid Disorders

One of the first tests developed in nuclear medicine was the assessment of thyroid function. The thyroid is a gland situated in the front of the neck, small in size, but big in importance. It manufactures an important hormone called thyroxin. Individuals in which the thyroid is overactive tend to be nervous, excitable, very active and extremely thin. An underactive thyroid,

FIGURE 6.23. The concept of using radioactive isotopes as tracers in medicine was not fully exploited until the invention of the cyclotron in 1931. Its inventor, Ernest O. Lawrence, is seen here with his second cyclotron in 1934. Many short-lived isotopes are made with a device of this sort. (Courtesy of the University of California Lawrence Berkeley Laboratory.)

on the other hand, leads to sluggishness, tiredness, depression and frequently a problem of excess weight.

None of these symptoms are unique to thyroid disorders and could equally well result from other causes. It is important, therefore, to have a simple test to identify those instances in which the thyroid is to blame, since these can be readily treated and in most cases cured. The procedure is as follows:

The patient drinks a small quantity of iodine, which is one of the essential ingredients used by the thyroid to make thyroxine. Many years ago, people who lived in mountainous areas far from the sea often had large goiters because they lacked iodine in their diet. The iodine used in nuclear medicine contains a tiny trace of an isotope of iodine which is *radioactive*, made from materials exposed in a nuclear reactor. An atom of radioactive iodine differs from ordinary iodine in having several extra neutrons in its nucleus, as a result of which it gives off both beta and gamma rays. Both isotopes of iodine, the normal and the radioactive, have identical chemical properties and are metabolized in the body in precisely the same way; most is excreted in the urine, but some is absorbed into the bloodstream, and finds its way to the thyroid gland. A gland that is overactive and producing too much thyroxine will collect a lot of iodine, which it needs to make the hormone. Conversely, an underactive gland will not need much iodine.

Twenty-four hours after giving the dose of iodine, a measurement is made over the patient's thyroid with a sensitive counter. The amount of iodine localized there is measured by the radiation it gives off. The normal range is known from previous tests on many normal volunteers. A low reading indicates an underactive thyroid, a high reading an overactive gland. This is usually the first simple test made; more elaborate investigations can follow as required.

The detectors used to measure radiation are so sensitive that these tests can be performed with incredibly small amounts of radioactive iodine. In a typical thyroid function test, the radiation dose to the gland itself is about 50 milligray (5 rad), while the dose to the rest of the body, in particular the sex organs, is not more than 0.1 milligray (10 millirads). The patient is subjected to a minimum of inconvenience, and no discomfort or trauma. No surgery is involved, or even the use of a hypodermic syringe. The iodine is taken by mouth, pleasantly flavored to make it palatable, and all subsequent measurements are made from outside of the body, with no physical contact.

When a patient, suspected of having an overactive or underactive thyroid, first consults an endocrinologist, the initial test used to assess thyroid hormone levels is likely to involve radioimmunoassay—a technique discussed later. The elegance of this technique lies in the fact that hormone levels are assessed from a blood sample, with no radioactivity given to the patient. However, a definitive measurement of thyroid function requires the administration of a dose of radioactive iodine, and the uptake 24 hours later. The activity of iodine used can be as small as 185 Becquerel (5 microcurie).

In a significant proportion of the patients diagnosed as having an overactive thyroid, the treatment of choice is again radioactive iodine. This time a dose is given which is a thousand times larger than that used to test the function of the thyroid in the first place. Most of the radioactive iodine is concentrated and localized in the thyroid, subjecting the gland to intense irradiation and reducing its activity to a normal level. The symptoms are relieved soon afterwards. The radiation dose to the thyroid may be as high as 100 Gy (10,000 rads), and the average dose to the whole body from radioactive iodine in the circulating blood is about 0.14 Gy (14 rads). This is a large dose of radiation, and can be justified only if there are substantial benefits to offset the risks. Eight out of 10 of the patients treated in this way are relieved of their symptoms and are able to live a normal life; this is a higher rate of success than the surgeon can claim and at a smaller risk. The treatment is inexpensive to both patient and hospital and involves no more than drinking a glass of an iodine solution.

If the isotope treatment is *not* used, the alternative is an operation to remove part of the gland surgically. This carries with it a risk, very small but not negligible, of death on the operating table, a scar on the neck, and a poorer chance of success in relieving the symptoms.

Many thousands of thyroid patients have been treated with radioactive iodine and have been followed carefully for a period of years. As yet, no deleterious effects can be detected which can be attributed to this form of treatment. It is virtually certain that if a larger group of patients could be studied, amounting to several millions, then a small incidence would be observed of leukemia, other forms of cancer, or genetic anomalies in the offspring resulting from the substantial doses of radiation involved. However, none can be seen in over 30,000 patients treated, so the risk must be small indeed, and acceptable in view of the success of the treatment and the greater hazards involved in alternative procedures.

Imaging

After administration of a material which carries with it a radioactive label, it is useful in some cases to study the pattern of distribution of this material in the body. This is obtained by using a "gamma camera," similar in principle to

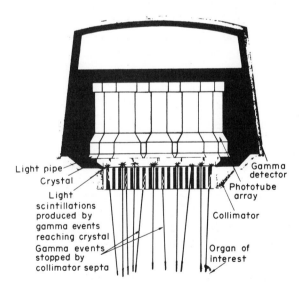

FIGURE 6.24. Diagram to illustrate the principle of a gamma camera. Gamma radiation from the radioactive isotope in the organ of interest within the patient falls on a sensitive crystal; there is a collimator to accept only gamma rays perpendicular to the crystal, so that the pattern on the crystal reflects the distribution of activity in the patient. The crystal converts the gamma rays to light which passes up the light pipes and is viewed on a television screen (courtesy of Nuclear Chicago Corporation).

an ordinary camera, except that the image is produced by gamma rays falling on a sensitive crystal instead of light rays on a film. The principle is illustrated in Figure 6.24. Gamma rays from a radioactive isotope distributed within the patient's body are collimated by small holes in a thick lead screen (comparable to the aperture in a conventional camera) and form an image on a large sensitive crystal which glows when radiation falls on it. The development of gamma cameras was contingent upon the availability of these huge crystals of sodium iodide, up to 12 inches in diameter, which are difficult and expensive

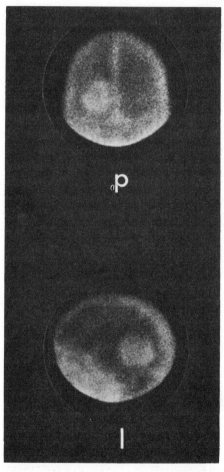

FIGURE 6.25. Picture taken with gamma camera of a patient who had received radioactive technetium in the form of pertechnetate. A brain tumor is clearly visible on the posterior view (marked P) and the left lateral view (marked L). Increased activity is present in the tumor because of a disruption in the blood-brain barrier allowing the isotope to enter. (Courtesy of Dr. Philip Johnson, Columbia University, New York.)

to fabricate. The image formed on the crystal can be photographed with a Polaroid camera and accurately records the pattern of isotope distribution within the body.

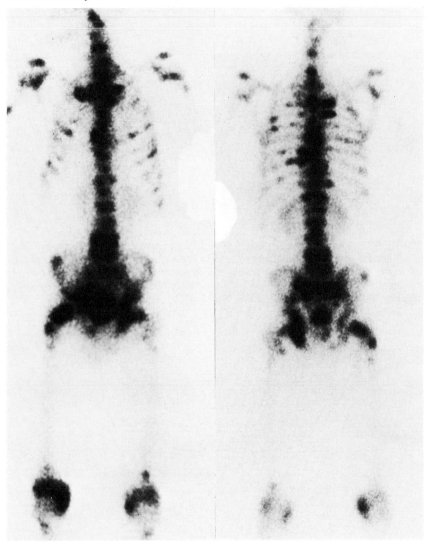

FIGURE 6.26. Gamma camera "image" of a patient who had received radioactive technetium in the form of diphosphonate. This material is taken up by the bones. The right and left-hand pictures are views from the back and front, respectively. The scintigram shows multiple areas of abnormally increased uptake of radioactivity, corresponding to many secondary tumors in the bones, particularly in the spine, spread from cancer of the prostate. (Courtesy of Drs. Philip Alderson and Barbara Binkert, Columbia-Presbyterian Medical Center, New York.)

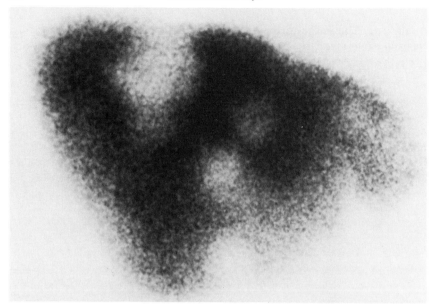

FIGURE 6.27. Gamma camera "image" of the liver of a patient who had received sulfur colloid labeled with technetium 99m. The light areas correspond to less uptake of the radioactive material and suggest metastatic disease, i.e., secondary cancer with a primary tumor elsewhere. (Courtesy of Drs. Philip Alderson and Barbara Binkert, Columbia-Presbyterian Medical Center, New York.)

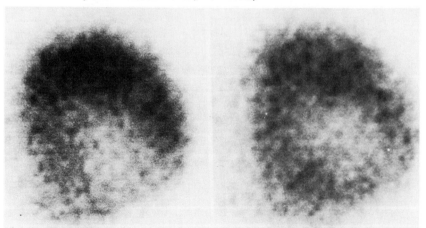

FIGURE 6.28. Gamma camera images of the heart of a patient who had received radioactive thallium-201. The picture on the left was taken during exercise. The right at rest. The region at the bottom left shows decreased uptake during exercise, but the same region appears normal at rest. This finding indicates ischemia in the inferior myocardium—i.e., diminished blood flow to part of the muscle of the heart. (Courtesy of Drs. Philip Alderson and Barbara Binkert, Columbia-Presbyterian Medical Center, New York.)

Gamma cameras are used for many purposes. A brain tumor can often be detected and its position located by using a gamma camera on a patient who has been given radioactive technetium (Fig. 6.26). The activity is *higher* over the tumor because the blood-brain barrier has been perturbed. A clear example is shown in the accompanying pictures. Some compounds, such as diphosphonate, are taken up in the bones, and if labeled with a radioactive material such as technetium, the details of the skeleton can be seen. Abnormalities, including tumors, are readily visible (Fig. 6.27). Metastatic cancer in the liver is detectable with nuclear medicine techniques (Figure 6.27), also abnormalities in the heart by studying images taken at rest and during exercise (Figure 6.28).

Positron Emission Tomography

Most radioactive materials used in nuclear medicine are produced in a reactor and decay by the emission of gamma rays which are given off randomly in all directions; to use the technical term, the emission is isotropic. (There may also be alpha and/or beta rays, but they cannot be detected outside of the body.) This represents one limit to the accuracy and sharpness of scans that can be obtained with conventional nuclear medicine techniques. The edge of an organ that has taken up a radioactive isotope such as, for example, the liver or kidney, is of necessity blurred because the radiation is given off in all directions.

There is a class of radioactive isotopes which decay by the emission of a positron, which is a particle having the same size as an electron but carrying a positive charge. In general, such isotopes are artificial i.e., man-made, and must be produced in a cyclotron. There are three important points—which constitute the good news and the bad news. *First*, positron emitting isotopes include Carbon-II, Nitrogen-13 and Oxygen-15—i.e., radioactive isotopes of elements that make up the tissues of the body. Convenient gamma-emitting isotopes of these elements cannot be made in a reactor. The availability of positron emitting radioactive labeled biological materials opens up a whole new field in medicine and research. *Second*, most positron emitting isotopes decay rapidly, with a half-life of seconds or minutes. When administered to a patient, therefore, the radioactivity is gone quickly and the radiation dose is small. This is the good news. The bad news is that a $2,000,000 cyclotron is needed on hand to produce the radionuclides; because they decay so rapidly, they cannot be transported over large distances. For this reason, the technique is confined to a few large medical centers. *Third*, the way in which a positron decays makes possible a special kind of detection and imaging. A positron cannot exist free and at rest in our world, which is made up of protons and electrons; consequently, when the fast-moving positron emitted by a radioactive nuclide slows down, it combines with an electron and "annihilates," that is, disappears completely, while the mass of the electron and the positron is converted into two photons of gamma rays. In order to conserve the laws of physics, these two gamma rays are emitted in exactly

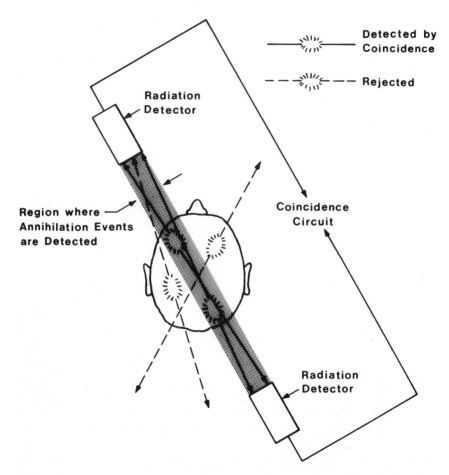

FIGURE 6.29. Illustrating the principle of Positron Emission Tomography (PET). The patient is given a radionuclide which emits positrons. The positron combines with an electron and "annihilates"—to form two 0.51 MeV photons which travel in exactly opposite directions. These are picked up by a radiation counter. A coincidence circuit is used to record only those photons detected simultaneously by the two opposing detectors—ensuring that only radiation from positron annihilation is recorded. (Courtesy of Dr. Michel Ter-Pogossian, Mallinckrodt Institute, Washington University, St. Louis, Mo.)

opposite directions, while the energy of each photon is precisely 0.51 million electron volts. Therefore, what is done in patients who have been given a radioactive material that is a positron emitter, is to position sensitive detectors above and below the patient, and to program the detectors to record only particles that are received simultaneously by *both* detectors; in this way the annihilation radiation from the positron is detected. This is referred to as *coincidence counting*. An image is produced on the detector in front of the patient and another image on the detector behind the patient; one knows with certainty that the point at which the radiation was emitted within the patient lies precisely along a straight line between these two images. The principle of a PET scanner is thus very simple, utilizing the basic fact that the annihilation radiation when a positron disappears consists of two photons traveling in exactly opposite directions.

Translating this basic principle into a practical and workable scanner turns out to be a difficult and an expensive proposition. A number of experimental prototypes have been developed, and one of them is shown in Figure 6.30. An example of the clinical usefulness to demonstrate a myocardial infarct is

FIGURE 6.30. Photograph of the Super PETT I scanner at the Mallinckrodt Institute, St. Louis. The patient lies on the couch with his head, or other part of his body to be examined, within the ring. The ring contains the radiation detectors which pick up annihilation radiation from positrons emitted within the patient. The information is fed to a computer which produces an image of the radioactive material within the patient's cross-section. (Courtesy of Dr. Michel Ter-Pogossian, Mallinckrodt Institute, Washington University, St. Louis, Mo.)

shown in Figure 6.31. This exciting development is likely to be used much more widely in the future, but the exact role of PET scanners and the extent to which they will replace other imaging devices is not yet clear.

Measurement of Blood Volume

By using a simple but ingenious technique, radioactive isotopes can be used to measure the volume of a fluid which is not accessible to more direct measurement. The blood is a good example.

It is possible to measure the *plasma* volume, or the *red cell* volume, either of which may be useful to the physician according to the particular situation. To measure the plasma volume, a sample of human serum albumen is used which is labeled with radioactive iodine; for the red cell volume, a sample of

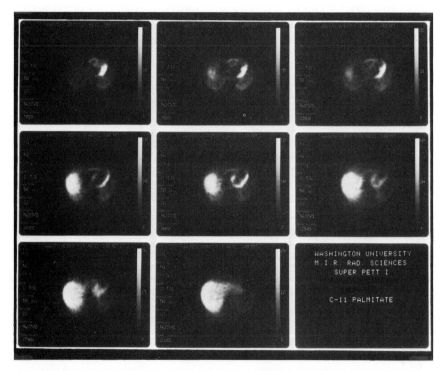

FIGURE 6.31. Series of pictures taken with Super PETT I of a patient suffering from a myocardial infarct. The patient received Carbon-II labeled palmitate and the series of pictures corresponds to "slices" through the chest at different levels. The top left is the highest level—through the heart—and shows up the infarct clearly as a bright spot. The bottom right picture is at the lowest level, which is the liver, which accumulates palmitate. (Courtesy of Dr. Michel Ter-Pogossian, Mallinckrodt Institute, Washington University, St. Louis, Mo.)

human red blood cells is labeled with radioactive chromium. These radiopharmaceuticals are now readily available from commercial suppliers.

In either case the labeled preparation, contained in a syringe, is measured with a Geiger counter or scintillation counter to assess the amount of radioactivity it contains. The contents of the syringe are then injected into a vein and time allowed for the blood to circulate and thoroughly mix the injected sample with the total blood volume. Following this, a sample of blood is withdrawn in a syringe and the amount of radioactivity it contains measured in the same counter. From the ratio of the total radioactivity injected to the small fraction that is recovered in the blood sample, it is a simple matter to calculate the total volume either of the plasma or of the red cells as required.

In fact, automated devices are now available which not only make the measurements of radioactivity, but also perform the calculations and display the result. This is useful, for instance, in an operating theatre, when immediate answers are needed as a guide to treatment in an accident case, where there has been a large blood loss.

The basic principle of this test is known as the dilution technique and has a wide range of uses, by no means confined to medicine. It is the simplest and most reliable way, for example, to measure the volume of a lake or the flow-rate of a river. Indeed, it can be applied to any problem that involves the estimation of a volume that cannot be measured directly, but to which radioactive samples can be added, mixed with the whole volume, and later recovered.

RADIOIMMUNOASSAY

Radioimmunoassay (RIA) is an ultrasensitive technique for measuring the tiny amounts of hormones in, for example, the blood or gastric juices of patients—levels too small to be assessed by any other means. Hormone excess or deficiency can be speedily and readily diagnosed in this way. The sensitivity of RIA is remarkable; for example, hormone levels of a picogram—i.e., a millionth of a millionth of a gram—can be measured readily. Early uses included the measurement of plasma insulin in man and the assessment of vitamin B_{12} levels, but RIA has since been applied to measure a dozen different hormones, including pituitary, pancreatic, chorionic, gastrointestinal, and thyroid hormones, as well as nonhormone substances such as serum proteins, enzymes, and tumor antigens.

The use of these powerful techniques has spread to the point where, in a typical large hospital, 20 to 30 thousand tests per year may be ordered. Indeed, they have become a relatively standard item, comparable to a chest x-ray or blood count, ordered by physicians in the routine workup of a patient. Their usefulness is beyond doubt, and most RIA tests are now performed

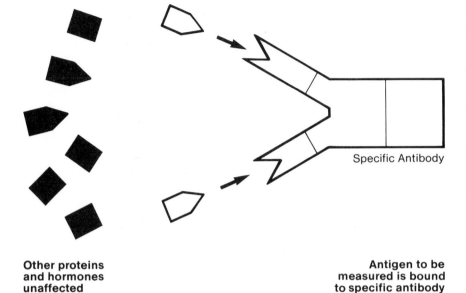

**Other proteins
and hormones
unaffected**

**Antigen to be
measured is bound
to specific antibody**

FIGURE 6.32. Illustrating the principle of radioimmunoassay. The basis of the technique is the incredible specificity; antigens are bound only to the specific antibody—all
others are ignored.

using commercially available, prepackaged kits. The majority of the hormone assay tests have long since passed out of the realm of the research
laboratory into routine use.

While endocrinology was the first home of RIA, the field has expanded to
include measurement of pharmacological agents, measurement of enzymes
and what may prove to be a big field—measurement of viruses. RIA of
hepatitis-associated antigen has become the method of choice for testing
infected blood in Red Cross and other blood banks in the United States,
where transfusion-transmitted hepatitis has been a significant public health
problem. RIA can be used to detect mouse leukemia virus with incredible
sensitivity so that it may become a new tool for studying the role of viruses in
cancer. Infectious diseases have become less prominent as causes of death in
industrialized countries, but remain a major health problem throughout the
rest of the world. Simple, inexpensive methods of identifying carriers of
disease would facilitate eradication of these diseases. Early detection, for
example, for the growth of tubercle bacilli may be possible.

Radioimmunoassay (RIA) is simple in principle. The concentration of the
unknown antigen from the patient is obtained by comparing its inhibiting
effect on the binding of radioactively labeled antigen to a specific antibody
with the inhibitory effect of known standards.

PRINCIPLE OF RADIOIMMUNOASSAY

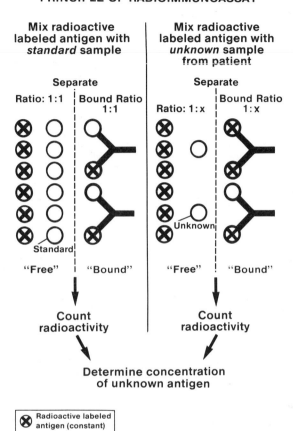

FIGURE 6.33. Illustrating the principle of radioimmunoassay. The unknown hormone sample from a patient is compared with a standard known sample of the same hormone as follows: — Radioactive labeled antigen is mixed with known and patient samples. The bound ratio will be the same as the free ratio. Bound material is separated from the free and the radioactivity counted. In this way the (for example) hormone level in a patient can be determined from the standard sample. The great sensitivity of radiation detectors makes this technique possible. The tiny amounts of hormones could never be detected much less measured by standard chemical methods.

The basic reaction of all immunoassays is the same (Figure 6.33) in that an antigen binds to an antibody. An antigen is a chemical which elicits an immune response—an antibody is the response. The antigen to be measured may be, for example, a hormone such as dioxin in the fluid of a patient under test. A radioactive labeled sample of the same antigen is mixed with the sample from the patient and with the appropriate antibody specific for that antigen. At the same time in a parallel test tube, the same radioactive labeled antigen and antibody are added to a standard sample of the type of antigen to be measured (left-hand side of the figure).

Radioactive labeled antigen competes with both standard and test samples of antigen for a limited number of antibody binding sites. This reaction will leave both labeled and sample antigen in two phases. Antigen bound to antibody is called the "bound" phase; the unbound remainder is called the "free" phase. Making use of the fact that the antibody-antigen complex is a larger molecule than the antigen itself, these phases are separated using one of several filtering techniques. The radioactivity in the bound phase is then counted in a gamma scintillation counter; the radioactive labels most often used are ^{125}Iodine or ^{57}Cobalt. By a comparison of the amount of radioactivity in the bound phase from standard and patient samples, the amount of hormone in the patient sample can be calculated.

One of the advantages of the technique is that the radioactive material is not administered to the patient, but added to samples of, for example, the patient's blood in a test tube. Consequently there is no risk or hazard posed to the patient by radiation, and the test may be repeated many times to monitor the success or failure of treatments designed to correct the hormone imbalance.

RADIATION THERAPY AND ONCOLOGY

The three major forms of cancer treatment are radiation therapy, surgery and chemotherapy. Radiation therapy and surgery are used principally for the control of local tumors, while chemotherapy is reserved for disseminated or widespread disease.

The use of radiation for the treatment of disease, especially cancer, was begun within a year or two of the discovery of x-rays. France, Sweden and later Great Britain were the early leaders in the field. The major contribution of the United States came during the years following the Second World War, by which time radiation therapy had won an accepted place in every large medical center.

Radiation therapy is based on the ability of x-rays (or any other form of ionizing radiation) to kill cells in the sense that they lose their ability to divide and to proliferate. Tumors grow because of uncontrolled division of the malignant cells, and so any agent which terminates this ability to divide controls the growth of the tumor. Radiation is a particularly effective agent in this respect. Successful treatment is based upon a number of factors, most important of which is careful direction of the beam and the delivery of a schedule of doses spread over a protracted period of time.

The Strategy of Treatment

The x-ray beams are accurately aimed to encompass the known tumor mass, while including as little normal tissue as possible. Critical and important normal tissues, such as the spinal cord or the eye, are avoided if at all possible. However much care is exercised, it is impossible to avoid irradiating some normal tissues. In the first place, the radiation must pass through

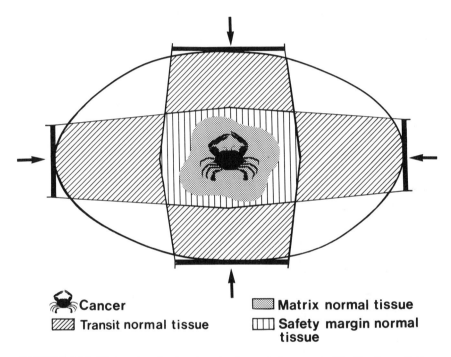

Cancer   Matrix normal tissue

Transit normal tissue   Safety margin normal tissue

FIGURE 6.34. Illustrating the treatment of cancer by x-rays. Ideally, the radiation should be given to the cancer alone and none to normal tissues, but this is not possible because in general the cancer lies deep within the normal tissue. The best that can be achieved is to maximize the dose of x-rays to the cancer and minimize the dose to normal tissue by using the cross-fire technique. The drawing is a cross-section of the human body through the pelvis with four beams of x-rays aimed to encompass the TARGET volume; this is made up of: (a) Overt malignant cells—the cancer; (b) Normal tissue which is intimately associated with the growing cancer and cannot be separated from it—this is called the matrix normal tissue. (c) A safety margin of about 1 to 2 cm around the cancer, to allow for the fact that even the most sophisticated diagnostic techniques cannot distinguish the tumor edge with great accuracy. The Target volume is treated with a high dose in an effort to sterilize the growing cancer. To reach a cancer deep within the body, the x-rays must pass through transit normal tissue. The target volume receives a higher dose than the transit normal tissue because in the cross-fire technique all four treatment beams overlap in this region, whereas any given portion of the transit normal tissue is treated by only one or two beams.

overlying normal tissues in order to reach a deep-seated tumor. In addition, the border of a tumor is difficult to delineate exactly, and it is prudent to irradiate a sufficiently large volume to avoid the likelihood of missing parts of the tumor which may be invading adjacent structures. This is illustrated in Figure 6.34.

The object of the treatment plan is to deliver a lethal dose of radiation to the tumor, while not producing too much damage to the surrounding normal tissues. The effect of radiation is not selective in tumor cells alone, but affects normal tissues as well. However, it has been found by experience that fractionated radiotherapy, in which a small dose is delivered five days a week over a period of four to eight weeks, is more effective in killing tumor cells while minimizing damage to normal tissues. This is a general rule for most, but not all, types of tumors. In the weeks that follow the completion of a course of treatment, the tumor cells which have been irradiated die while attempting to divide and are slowly broken down and removed by the body. The proliferation of normal cells must not only make good the damage caused to the normal tissues by the radiation, but must also grow in and repopulate the void left by the disappearance of the tumor.

THE CANCER PROBLEM

It is a commonly held belief that cancer is much more prevalent than it used to be. Everyone, it seems, has a friend or close relative who has died from cancer. There are two reasons for this. First, causes of disease and death are now more accurately diagnosed than they were a generation ago. Second, most people live longer nowadays, as the infectious diseases have been virtually eliminated as a cause of death in younger people; if you don't die at an early age from tuberculosis or diptheria, you are more likely to die of cancer at an advanced age. So to this extent it's true that cancer is more prevalent, in as much as more people live long enough to contract it. On the other hand, the likelihood of dying of cancer *at a given age* has not changed much since records have been kept. There are a few important exceptions. One is stomach cancer, which has decreased steadily and significantly in the industrialized nations ever since the 1930s, due probably to improved ways of food storage. The other is lung cancer, which has steadily increased and can be attributed almost entirely to smoking. At the present time, the United States has the highest incidence of lung cancer of any nation in the world—a direct consequence of the boom in smoking which occurred in World War II. Twenty years ago, lung cancer was more common in Great Britain than in the United States because, there, the big increase in smoking came in World War I. The third exception is death from cancer of the cervix in women, which has fallen steadily due to improved hygiene and early detection. With these three important exceptions, most common types of cancer have remained steady for half a century. Nor is it possible to attribute any significant proportion of cancer to the ills of modern society—air pollution, food additives, radiation, etc.

On the other hand, cancer is for the most part an avoidable disease, as is evident from the fact that the incidence of specific types of cancer varies so widely between different countries. For example, breast cancer is very common in the United States, but rare in Japan. The difference is not principally genetic since Japanese who move to the United States acquire the same breast cancer incidence as the natives within a generation. So it is clear that something in the American lifestyle causes breast cancer—but nothing new, it's been virtually unchanged for 50 years. The same is true of many other common types of cancer.

Despite all the money and effort put into cancer research, little has been actually proven of the causes of cancer. The one outstanding exception is smoking, which is clearly linked with 30 percent of all cancer deaths. By contrast, occupational factors cannot be involved in more than 4 percent and air pollution cannot account for more than 1 percent. Nuclear power is so far down the list that any estimate is probably meaningless, but it is certainly no more than 0.01 percent. It is a sad commentary on our society when outspoken protests are organized against nuclear power, while little is said about smoking which is at least a thousand times greater hazard.

The pattern and incidence of cancer varies with sex, age and location. Females are more prone to cancer than males, though the gap is closing rapidly because lung cancer is on the increase while cancer of the cervix (i.e., neck of the womb) is becoming less prevalent. In females, the common cancers are those of the breast, colon and uterus; in males the common lesions are of the lung, digestive tract and prostate. Figure 6.35 shows the distribution of cancers of various sites in men and in women. The figures were provided by the American Cancer Society and apply to the United States. Though statistics are similar for other industrialized nations, they differ markedly for the developing countries.

Cancer strikes at any age. With the exception of accidents, cancer kills more children than any other illness. In England, cancer is the leading cause of death in children 1 to 14 years of age, while in the United States about 4,500 children under 15 years of age die annually from leukemia, the lymphomas and cancer of the brain and kidney. Nevertheless, cancer strikes more frequently with advancing years so that by and large it is a disease of old age.

LOCALIZED AND ADVANCED CANCER

The great majority of cancers originate on the surface of some tissue or organ—such as the skin, the surface of the uterus, the lining of the mouth, stomach, intestines, bladder or bronchial tube, or the lining of a duct in the

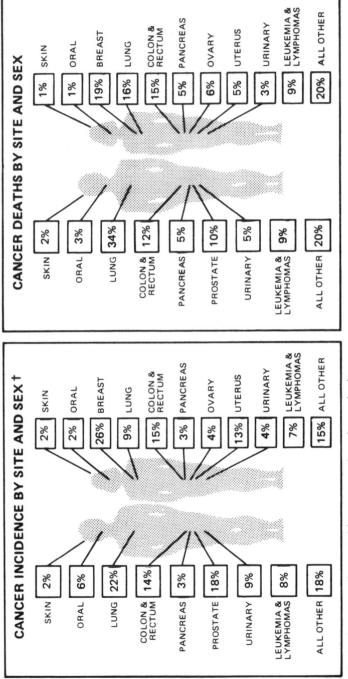

FIGURE 6.35. A comparison of cancer incidence and cancer deaths for various sites in men and women. In men, the most common cancers are of the lung, prostate and colon; in women, of the breast, colon and uterus. The difference between cancer incidence and cancer death reflects the varying cure rates for different types of cancer. The statistics were collected in the United States for 1983, but are similar in most Western industrialized countries. (Courtesy of the American Cancer Society.)

breast or prostate gland. For a time, such cancers remain in the lining or on the surface at the site of origin, following which they invade the underlying tissues. As long as the living cancer cells remain where the disease started, the cancer is said to be "localized."

The more dangerous phases of cancer are the later ones. Some of the cancer cells become detached and are carried through the lymph channels or bloodstream to other parts of the body. The process is known as "metastasis." But the body has a protective mechanism. The detached cells may be trapped in a lymph node in the region of the original growth. This slows down the spread for a time, but if left untreated, the cancer cells spread to distant parts of the body. This is *advanced cancer* and death is almost inevitable, though it may be delayed for a long time.

CURE AND PALLIATION

Radiation therapy has increased in importance in recent years to such an extent that about half of all cancer patients in the United States receive radiation at some stage of their disease. However, it is not considered, even by its most ardent advocates, to be the cure-all for cancer. Its usefulness is restricted to certain types of tumors, but within limitations it has proved its efficacy and won an accepted place in medicine. For a broad range of localized cancers, it provides a high probability of permanent cure, and is unquestionably the treatment of choice. Not only does it equal or excel the cure-rates which can be obtained by surgery, but the invisible rays are often less disfiguring and result in a much more acceptable cosmetic result. In advanced cancer, where a cure by any means would constitute a miracle, radiation therapy can still buy time and give relief from pain. This is called *palliation*, and while not as dramatic or as desirable as a cure, it is nevertheless of enormous value and a great blessing in terms of the relief of human suffering.

The current situation in the United States is shown in Figure 6.36 which illustrates the so-called $\frac{1}{2} \times \frac{1}{2} \times \frac{1}{2}$ rule, which emerged from a study of the patterns of care in hospitals across the nation. Of all cancer patients about $\frac{1}{2}$ receive radiation therapy at some stage in the management of their disease. Of those treated, about $\frac{1}{2}$ are treated with intent to cure, of which about $\frac{1}{2}$ are cured. Overall, then, $\frac{1}{8}$ of all cancer patients in the United States are *cured* by radiation therapy, which amounts to over 100,000 patients per year. About $\frac{3}{8}$ receive palliation. This is already a substantial achievement, but highlights how much room is left for further improvement.

AREAS OF SUCCESS

A legacy exists in the minds of most people of an association between radiation therapy and incurable disease. Many believe that if you are scheduled for radiation therapy, you are too far gone to be saved; this is simply not true.

Patterns of Care Study

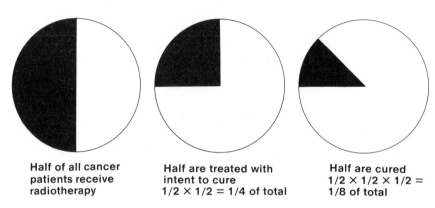

Half of all cancer patients receive radiotherapy	Half are treated with intent to cure $1/2 \times 1/2 = 1/4$ of total	Half are cured $1/2 \times 1/2 \times 1/2 = 1/8$ of total

FIGURE 6.36. According to the "Patterns of Care" study in the United States, about 800,000 new cases of cancer are diagnosed each year. Of these, HALF receive radiation at some stage in the management of their disease. Of those treated, HALF are treated aggressively with intent to cure, the others receiving palliation only because cure is not feasible. Of those treated with intent to cure, HALF are cured. Thus, 1/8, or 100,000 cancer patients, are cured each year by the use of radiation. This is already a significant achievement, but clearly there is much room for improvement.

Radiotherapy has evolved into an extremely precise and effective technology. The treatment by radiation of 10 different kinds of cancer—breast, cervix, larynx, prostate, uterus, bladder, testicle, tongue, floor of mouth, and Hodgkins disease—results in cure rates that equal or surpass surgery while preserving an intact body and organ function. Some of these examples require further comment.

Cancer of the uterine cervix has been a major cause of death among women, but with proper treatment many cases can be cured. There are two principal methods of using radiation to treat this type of cancer. Radiation may be beamed to the cancerous tissue from an x-ray or cobalt-60 therapy machine, or alternatively, radium sources may be placed directly into the uterus and vagina to produce intense local irradiation. Often, a combination of both methods is used, which is highly effective, particularly in early cancers.

Radiation is the treatment of choice for some cancers of the head and neck. The success rate is high, while normal or nearly normal function and appearance is preserved. Cancer of the larynx is a good example. When disease is confined to the vocal chords, 90 percent of the patients can be cured by radiation and retain their natural voice; by contrast, if surgery is performed the patient is at best hoarse, if he doesn't lose his voice altogether.

Five-Year Cancer Survival Rates*
for Selected Sites by Race, 1964-1973

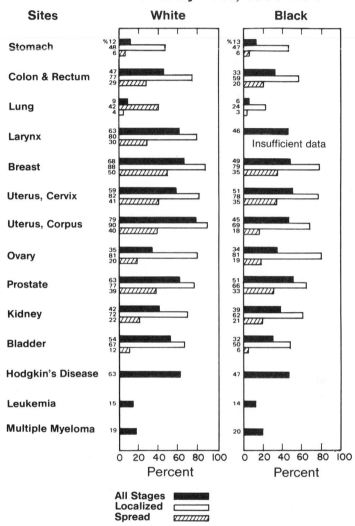

FIGURE 6.37. Survival rates for patients with various types of cancers. For many common cancers, the 5-year survival figures are excellent if they are diagnosed and treated early. In most cases, 5-year survival is essentially a cure. Early cases of cancer of the larynx, uterus, prostate and breast show a 5-year survival in excess of 75%. By contrast, they are poor for lung. (Courtesy of End Results Group, National Cancer Institute.)

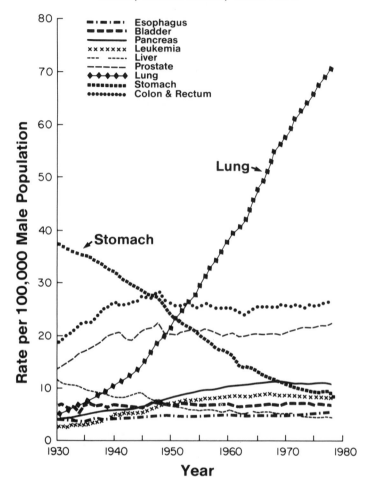

Age-Adusted Cancer Death Rates* for Selected Sites
Males, United States, 1930-1978

*Adjusted to the age distribution of the 1970 U.S. Census Population.
Sources of Data: U.S. National Center for Health Statistics and U.S. Bureau of
the Census.

FIGURE 6.38. Death rate from common types of cancer for men over a 50-year period from 1930 to 1978. Cancer of the lung has increased strikingly with the increase in smoking, while cancer of the stomach has decreased substantially as more effective methods of food preservation have been introduced. With these two exceptions, both of which can be accounted for, death rates from the majority of cancers have not altered over the past half century, which does not support the notion that cancer today is due to food additives, air pollution and all the other ills of today's society.

Supervoltage radiation therapy has proved to be extremely effective combined with chemotherapy in the treatment of Hodgkin's disease, a form of cancer which primarily effects the lymph nodes and lymphoid tissues. Certain bone tumors are curable by radiation, without the necessity for surgical amputation. The most promising new results concern Ewing's sarcoma, a usually fatal form of bone cancer which occurs among children and young adults. As techniques have improved, the role of radiation therapy has expanded in the treatment of many types of central nervous system cancers.

Age-Adjusted Cancer Death Rates* for Selected Sites
Females, United States, 1930-1978

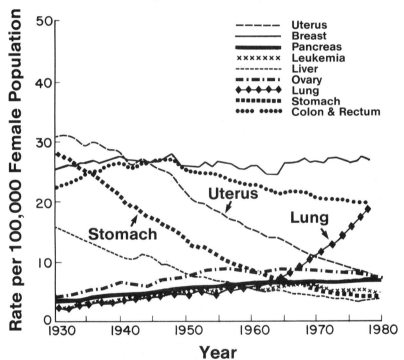

*Adjusted to the age distribution of the 1970 U.S. Census Population.
Sources of Data: U.S. National Center for Health Statistics and U.S. Bureau of the Census.

FIGURE 6.39. Death rate from common types of cancer in women over a 50-year period from 1930 to 1978. Cancer of the stomach and uterus have declined dramatically, while there is a sharp rise in deaths due to cancer of the lung since 1970, reflecting the increase in smoking in women. Death from the majority of other cancers has not altered much for over half a century.

For example, it is the only form of therapy which, at present, is capable of curing medulloblastoma, a form of brain tumor most common in children.

Certain germ cell tumors are highly sensitive to radiation, and some are readily curable. As many as 90 percent of patients with seminoma, a form of cancer of the testicle, may be cured by a combination of surgery and radiation therapy.

Radiation combined with chemotherapy has helped arrest retinoblastoma, a cancer of the eye, and preserve useful vision in children whose prospects for sight were otherwise unfavorable.

Skin cancer is extremely common, especially in the American Southwest and Australia, where white people of European descent are exposed to the intense ultraviolet rays from the sun. Fortunately, it is 95 percent curable by either radiation or surgery. For many patients, radiation therapy is the treatment of choice because it destroys the cancerous tissue with little damage to surrounding areas, so producing a superior cosmetic result.

TREATMENT MACHINES

The treatment machines commonly used today produce beams of x-rays or gamma rays, which are carefully and accurately aimed at the tumor. X-rays are generated in large electrical machines. In such devices, electrons are accelerated to high energy and acquire almost the speed of light, before hitting a target made of tungsten or gold. The fast electrons stop abruptly, and their energy is converted to x-rays. The higher the voltage of the machine, the more penetrating the x-rays produced. A certain amount of penetration is essential, otherwise it is not possible to deliver a sufficient dose of x-rays to tumors located deep within the human body.

In the 1930s, 200,000 volts was regarded as a high voltage, and such machines were described as "deep x-ray therapy" machines. Millions of cancer patients were treated with such machines, and some good results were obtained. In many cases, however, the penetration of the beam was inadequate and constant efforts were made to manufacture new machines with higher and higher voltages, limited always by the technology of the day. Since the Second World War, a new generation of treatment machines has emerged, based largely on engineering advances which were a spin-off of the war effort.

Linear accelerators owe their development to a technological breakthrough in England during the early days of the Second World War. The electronic oscillators which made radar possible, and guided the Spitfires to their targets during the Battle of Britain, now power cancer treatment machines all over the world. Most major radiotherapy centers have at least one linear accelerator, operating at about 4 to 18 million volts. Such machines are able to penetrate to the most deep-seated tumors.

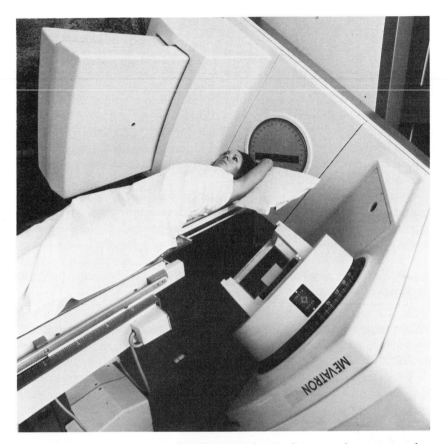

FIGURE 6.40. Illustrating a modern linear accelerator for generating x-rays used to treat cancer. The accelerating voltage is 6 million volts. (Courtesy of Siemens Medical Systems.)

Cobalt therapy units are an alternative source of penetrating radiation. These devices contain a large source of radioactive cobalt, which constantly gives off high-energy gamma rays. As previously explained, x- and gamma rays do not differ in properties or in the biological effects which they produce. Their distinctive names simply reflect the different ways in which they are produced. Cobalt is an element which occurs abundantly in nature; it is a metal added in small quantities to harden steel. It is not naturally radioactive, but if a sample is placed in a nuclear reactor and bombarded intensely with neutrons, some of the atoms gain extra neutrons and so form a new and unstable isotope of cobalt. In attempting to become stable again, these "radioactive" atoms give off gamma rays.

Radioactive cobalt, or cobalt-60 as it is called, is used as a source of gamma rays in treatment units. The radioactive source is contained in a large sphere of lead, about 3 feet in diameter, which protects the doctors and nurses from continual exposure to the rays. The patient is carefully aligned in front of the unit, and gamma rays emerge from a hole in the lead sphere, which is adjusted to be of a suitable size and shape. There are only a handful of reactors in the world which are powerful enough to produce cobalt-60 suitable for these purposes. Isotope production is a useful and peaceful application of the reactors which were initially developed to make atom bombs. The United States Department of Energy and Atomic Energy of Canada share the production of almost all of the radioactive cobalt-sources used in medicine throughout the free world.

Cobalt units have at least two advantages over linear accelerators; they require only a modest electrical supply, and very infrequent servicing by expert engineers. Consequently, cobalt units are particularly suitable for use in developing countries and in smaller community hospitals all over the world. Linear accelerators are more complex and sophisticated, and are usually found in large medical centers which have ready access to skilled physicists and engineers.

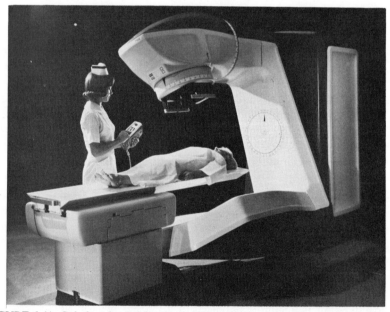

FIGURE 6.41. Cobalt unit used for the radiation therapy of cancer. With the nurse in the room, the radioactive source is in the "safe" position at the center of the treatment head which is essentially a sphere of lead. When the patient is in position, the nurse or x-ray technician leaves the room and the source is moved to the "on" position. (Courtesy of Atomic Energy of Canada Limited.)

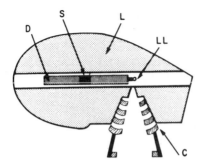

FIGURE 6.42. Illustrating the principle of a cobalt-60 therapy unit. A large source of radioactive cobalt (S) is housed in a sliding drawer (D). It is surrounded by a mass of lead (L). As shown, the source is in the "off" or safe position. A localizing light at the end of the source drawer shines down the collimator (C) and shows the size of the area that will be treated. When the machine is turned "on," the source drawer moves until the source is opposite the collimator and occupies the position LL in the picture. Gamma rays then pass out of the collimator and treat the patient. (Courtesy of Dr. Paul Goodwin and Harper and Row.)

IMPLANTS WITH RADIOACTIVE SOURCES

Instead of beaming x- or γ-rays into a patient from an outside source, an alternative strategy is to implant the tumor directly with radioactive sources. This was an early development, and implants have been used in cancer therapy from the turn of the century. Until the Second World War, the sources consisted of rigid needles containing radium, or small seeds containing radon gas. The disadvantage of radium is that the sources are thick and rigid and also relatively expensive since there is a limited inventory of radium in the world. The disadvantage of radon is the short half-life. It decays away in a few days and must be used soon after it is made.

Since World War II, a vast variety of artificial radioactive materials have been produced, and from this wide choice two have found a long-term place. (1) Wires of radioactive ^{192}Iridium. These flexible sources can be implanted into a tumor and left 3 to 10 days to give the required dose. Because of their flexibility, they are more comfortable for the patient than radium needles especially for implants in the head and neck. (2) Seeds of ^{125}Iodine. This isotope has a relatively short half-life of 60 days, and once implanted into a patient, the seeds are left permanently to decay away. The advantage is that a second operation to remove the seeds is not required, while the soft emission from ^{125}I does not pose radiation protection problems to other people. Such seeds are widely used, for example, for cancer of the prostate.

Implants with radioactive sources are not feasible for the bulk of human cancer; only about 5 percent, or 1 in 20 patients, are suitable. An implant is useful only so long as the tumor is localized and then only if it is accessible. But in those cases, it is the treatment of choice. Examples include early cancer of the tongue, floor of mouth, vocal cord and breast. If treatment with an implant is feasible, the result is control of the local tumor together with preservation of normal organ function. The advantage is obvious with cancer of the tongue, since the alternative to radiation therapy is removal of the tongue by surgery, a very traumatic procedure. In summary, then, radioactive implants represent an important choice for a small sector of the cancer population. Their use is illustrated further in the section on breast cancer.

BREAST CANCER

No disease provokes more fear and dread among women, or more controversy over how best to treat it, than breast cancer. In Western industrialized countries, about one out of 11 women can expect to develop breast cancer, and the incidence appears to be rising, especially among younger women. In the United States alone, this translates into about 111,000 new cases that will be diagnosed each year, while about 37,000 will die of the disease.

The traditional treatment for breast cancer is surgery, combined often with radiation or chemotherapy, or both. But how drastic the surgery should be has long been the subject of serious debate. In the earlier days of medicine the surgeon's answer was to remove the entire breast, which not only disfigures the woman physically but leaves psychological scars as well. The rationale was that anything less drastic might leave traces of the malignancy that would regrow and spread. Using this reasoning, the seventeenth century Italian surgeon Marcus Aurelius Severino extended the technique by cutting away an enlarged underarm lymphnode in addition to the breast. An even more drastic form of the operation, the radical mastectomy, was developed a century ago by a Johns Hopkins University surgeon by the name of William Stewart Halstead. Designed to minimize the chance of leaving behind any malignant cells, this procedure involves removal of not only the entire breast but also the two layers of chest muscles underneath and the lymph nodes in the armpit. This mutilating operation limits the movement of the patient's arm because of the removal of the chest wall muscles and leaves the arm permanently swollen because of the removal of the lymph glands. In advanced cases of breast cancer where the cancer has spread to the lymph nodes and invaded to a point where it is attached to the underlying muscle, the Halstead radical mastectomy is probably necessary. But are radical mastectomies always necessary to give the patient a best chance of cancer-free survival? Many doctors think not, especially when the cancer has been caught in the early stages; as surgery becomes more limited or is dropped altogether, the role of radiation therapy becomes central.

In a number of hospitals in the United States, on both East and West Coasts, very early small breast cancers are treated entirely by radiation. The cancer itself is implanted with wires of radioactive Iridium-192, which stay in place for 3 to 7 days, and give an intense localized dose of gamma-rays. The whole of the affected breast is subsequently given a more modest dose of radiation by external beam therapy. The cosmetic results are so good that 6 months to a year later it takes a keen eye to distinguish the treated from the untreated breast. A long-term study of hundreds of women treated in this way, compared with those receiving extensive surgery, shows no significant difference in the survival rate or in recurrence of disease. For cases involving a larger, but still early, breast cancer, external beam radiotherapy is added after conservative surgery to remove just the tumor itself rather than the whole breast—the so-called "lumpectomy." This procedure is only indicated where there is no evidence that the tumor has spread widely, but then the results of this combined treatment are found to equal those of more extensive surgery. In women with early breast cancer, many doctors now believe that a radical mastectomy involves unnecessary mutilation and does not improve the chances of long-term, cancer-free survival, compared with a more conservative treatment that is cosmetically and psychologically more satisfactory. The woman with early breast cancer may be well advised to look around and seek a second opinion before agreeing to extensive surgery; it is her body and she should make the choice!

PROGRESS IN TREATMENT

The introduction of high-voltage machines, such as linear accelerators and cobalt units, made possible a dramatic improvement in treatment techniques. Because of the more penetrating radiation, and the greater flexibility of these modern machines, it became possible to deliver larger doses of radiation to the known tumor volume and smaller doses to surrounding normal tissues. These technical innovations have been partly responsible for the dramatic improvement in cure rates during the post-war years.

However, success depends on far more than technical advances and developments in machine design. Great strides have been made in training specialized physicians who devote their entire efforts to the use of x-rays and other ionizing radiations for the treatment of disease. A specialist in this field is known as a radiotherapist or radiation oncologist. His patients are almost all people with cancer, and he sees a remarkably broad spectrum of malignant diseases: cancers of the skin, lip, oral cavity and pharynx, paranasal sinuses, lung, esophagus, cervix and corpus uteri, ovary, testis, urinary bladder, prostate, colon and rectum, brain and pituitary tumors, Hodgkin's disease and other malignant lymphomas, and even some of the leukemias.

In major medical centers, the cancer patient is seen first at a joint clinic attended by surgeons, radiotherapists and chemotherapists, so that the best available form of treatment is chosen in consultation. By working together, physicians with diverse specialized knowledge learn to know and respect the strengths and weaknesses of their colleagues. The patient benefits from the availability of all treatment alternatives. For some types of cancer, surgery is clearly the best—for others, radiation therapy is the treatment of choice, whereas in other instances chemotherapy is most likely to be successful. This decision must be made at the outset when the patient is first seen at a joint clinic, with no consideration other than the well-being of the patient.

Modern radiation therapy in the leading centers is truly a team effort, led by the physician, but ably supported by representatives of such basic sciences as physics, biology and chemistry. Like any other branch of medicine, radiation therapy is partly an art and partly a science, and as a consequence much depends upon the expertise and clinical judgment of the physician. At the same time, many of the most significant advances that have been made are a direct result of active collaboration with the basic sciences. The physicist has been largely responsible for better methods of radiation dose measurement and the development of penetrating megavoltage radiation that can reach deep-seated tumors. The biologist, by means of laboratory experiments, has contributed to a better understanding of the effects of radiation on tumors and especially on normal tissues, which has consequently led to modifications in the strategy of treatment. The chemist has been active in producing new types of pharmaceuticals which can be used in conjunction with radiation to increase its effectiveness in controlling the growth of tumors.

NEW DEVELOPMENTS

One of the most exciting new developments in radiation therapy is the use of neutrons for the treatment of cancer, in place of x- or gamma rays. At present, the studies are new and experimental, but early results are encouraging. This is a particularly exciting development because it is the direct result of laboratory research by physicists and biologists, in close collaboration with their medical colleagues.

The idea came about as follows. A cancer is a collection of cells growing out of control. It is malignant, that is, it will kill, because it grows without regard for the welfare of the host. As a result of the disorganized growth, the pattern of blood vessels is often inadequate, and consequently, parts of the tumor suffer from a lack of nutrition. It is characteristic of tumors as they grow larger that areas of necrosis appear, where cells are dead or dying. Small numbers of cells at the border of these areas of necrosis are still alive, but are deficient in supplies of all sorts of nutrients, including oxygen. At first sight it might be thought that such cells, laboring under a nutritional disadvantage, would be easier to kill with x-rays. In fact this is not the case. The cell-killing effect of x-rays is strongly dependent on the availability of oxygen; cells which are short of oxygen are to a large extent *protected* from the effects of x-rays. This has been proved convincingly in the laboratory, and it is thought that in the human too small a number of cells which are deficient in oxygen may be able to survive the radiation treatment and form a nucleus for the regrowth of the tumor. The point of using neutrons is that their cell-killing effect is *less* dependent on the availability of oxygen than is the case for x-rays. Oxygen-deficient cells are spared to a lesser extent, and the regrowth of the tumor from surviving cells is made less likely. Of course, not all tumors would be expected to benefit from treatment with neutrons, only those large tumors containing oxygen-deficient cells.

Neutrons are uncharged particles, and constitute one of the basic building bricks of which all atoms, except hydrogen, are composed. To be useful for

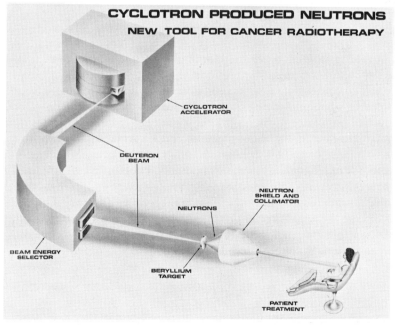

FIGURE 6.43. Schematic diagram to show how a large cyclotron is used to produce a beam of deuterons which are guided and shaped to hit a beryllium target to product neutrons, used for cancer therapy. (Courtesy of the Naval Research Laboratory, Washington, D.C.)

cancer treatment, they have to be accelerated to a speed close to that of light and given an energy corresponding to many millions of volts. This requires equipment vastly more complex and expensive than is required to produce x- or gamma rays.

Therapy with neutrons has been actively under way at the Hammersmith Hospital in London since 1967. A 16-million-volt cyclotron accelerator is used to generate the neutrons, and several hundred patients have been treated in this way. This machine is suitable for treating only tumors in the region of the head and neck because of the limited penetration of the beam. Consequently a new higher energy machine (60 MeV) is being constructed in the U.K. near Liverpool. In the United States, the initial experience of neutron therapy was with large atom-smashing machines, built several decades ago for high-energy physics research and converted to treat cancer patients. They all suffered from the disadvantage that they were not housed in or near a major hospital. Consequently the National Cancer Institute in the United States has funded the building of four new hospital-based neutron generators to test their usefulness in cancer therapy. They are to be located in Houston, Seattle, Philadelphia and Los Angeles. Similar programs are also underway in Japan and in Germany.

PROTONS

At the Massachusetts General Hospital in Boston, a beam of protons is in use for cancer treatment. By choosing the energy of the proton beam, it is possible to concentrate the radiation dose into a limited volume surrounding the tumor while minimizing dose to the surrounding tissue. This offers the attractive possibility of increasing the killing effect in the tumor without a similar increase in the damaging effects to normal vital structures. For example, this beam has been used with great success to treat choroidal melanoma of the eye—a relatively rare tumor, but one that is difficult to cure by any other means without loss of the eye. The sharp cut-off in dose outside the treatment volume that can be achieved so well with protons is particularly valuable in the eye. Whether or not protons will find a wider application in radiation therapy remains to be seen, but they already have demonstrated their usefulness in a specialized situation.

HEAVY IONS

At the Lawrence Berkeley Laboratory of the University of California, high-energy heavy ions are being used experimentally for the treatment of cancer. These are nuclei of common elements such as carbon, neon, argon, or silicon.

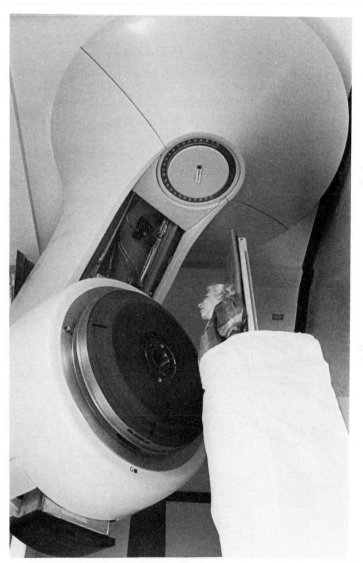

FIGURE 6.44. Photograph of the high-energy neutron therapy facility installed at the M.D. Anderson Hospital and Tumor Institute, Houston, Texas. The accelerator, built by Cyclotron Corporation, produces neutrons by bombarding a beryllium target with protons accelerated to 44 million volts. (Courtesy of Dr. Peter Almond and the M.D. Anderson Hospital and Tumor Institute.)

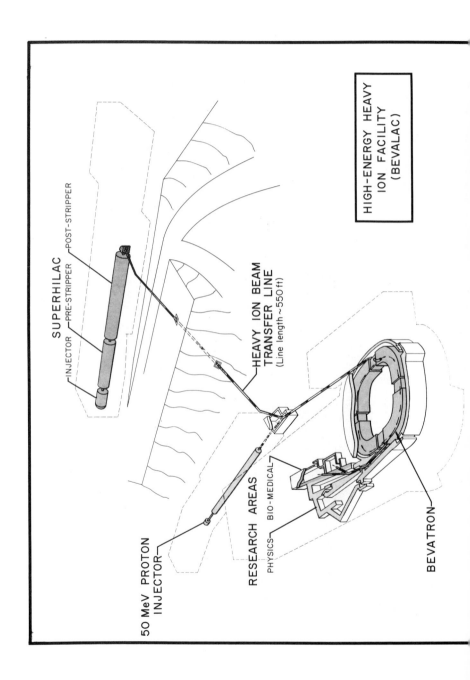

SUPERHILAC

INJECTOR — PRE-STRIPPER — POST-STRIPPER

HEAVY ION BEAM
TRANSFER LINE
(Line length ~550 ft)

HIGH-ENERGY HEAVY
ION FACILITY
(BEVALAC)

50 MeV PROTON
INJECTOR

RESEARCH AREAS

PHYSICS   BIO-MEDICAL

BEVATRON

FIGURE 6.45. (*Top*) Schematic of the link-up of the Super Hilac and the Bevatron to produce the Bevalac. This is the only facility in the world capable of accelerating heavy ions to high energies suitable for cancer therapy, and is located at the Lawrence Berkeley Laboratories of the University of California. (*Bottom*) Aerial photograph showing the layout of the Bevalac. Heavy ions are produced and initially accelerated in the Super Hilac and are then piped 500 ft. down the hillside and injected into the Bevatron to be accelerated to high energies. (Courtesy of the Lawrence Berkeley Laboratory, University of California.)

An elaborate and expensive facility is needed to generate these particles at the high energies needed for them to penetrate deep within the human body. An energy, for example, of many GeV* is needed for a particle to reach a depth of 10 cm. At present Berkeley is the only center in the world where such particles are used clinically, but there are plans for similar facilities to be built in Europe. The rationale of using high-energy heavy ions is to combine the physical advantages of protons with the biological advantages of neutrons— i.e., to exploit the property of a charged particle beam that it can be made to concentrate dose in the tumor volume, while at the same time the nature of the radiation is such that cell killing does not depend heavily on the availability of oxygen. This is a highly experimental form of treatment, and it will be many years before it can be evaluated and compared with more conventional methods of radiation therapy.

HYPERFRACTIONATION

It is clear that the exotic and expensive new treatment modalities, such as neutrons, protons and heavy ions, can be tested in only a limited number of centers. A strategy being tested in many more centers at the present time, especially in the United States, is *hyperfractionation*. In this regimen, two small treatment doses (each of about 1.1 to 1.2 Gray, i.e., 110 to 120 rads) are given each day, one in the morning and one in the afternoon, and continued 5 days/week for up to 8 weeks. This is, of course, expensive in terms of machinery and manpower and is difficult to implement in countries with a nationalized health service. However, evidence is mounting which supports the idea that a long protracted treatment leads to a sparing of normal tissue damage and excellent long-term cosmetic results.

RADIOACTIVE IMMUNOLOGY

The most elegant aspect of immunology is the exquisite specificity of the system. Foreign cells or molecules are "recognized" instantly and the defense system alerted. This characteristic is exploited in radioactive immunology, developed at Johns Hopkins Medical School and now used in a handful of other institutions in the United States. The basis of the technique is to make antibodies specific for the patient's tumor and use them to carry a radioactive isotope to the tumor in order to irradiate and destroy the tumor. Currently, radioactive Iodine-131 is the isotope used.

---

*1 GeV = 1000 million eV

Much effort has been devoted to manufacturing antibodies as a primary treatment of cancer, and this is an exciting long-term possibility, but one of the inherent weaknesses at present is that the antibodies may be effective against some, but not all, of the cells of the cancer. In the case of an isotopic immunoglobulin, the antibody carries the radioactive material and concentrates it in the *area* of the tumor so that the β and γ rays emitted by the radioactive Iodine-131 then destroy all of the cancer cells within a considerable range. The technique does not rely on the antibody homing in on each and every cancerous cell. The limiting factor of the technique, which is still very much in the development stage, is the production of suitable antibodies and the labeling of them with sufficient amounts of radioactive isotopes in such a way that the isotope remains firmly attached and does not float free. The most encouraging early results have been reported for the treatment of hepatomas, i.e., primary liver cancers. Ferritin is a naturally occurring antigen; it is also found to be synthesized and secreted by hepatoma cells. If a radiolabeled antibody (antiferritin I^{131}) is introduced in the patient, it binds to the ferritin and deposits I^{131} in the tumor site. One third of the patients suffering from a certain type of hepatoma, treated with antiferritin immunoglobulin carrying radioactive Iodine-131, are alive and well for over a year, when otherwise this disease inevitably leads to death within 3 months.

THE TANGIBLE BENEFITS

From this brief survey of the uses of radiation in medicine for diagnosis and for treatment, the significant benefits to the individual and to the community are at once obvious. Anyone who is ill and faces even minor discomfort, much less the threat of death, will not hesitate to seek medical help and is unlikely to refuse x-rays if they are advised by the physician. With very few exceptions, this is a wise and prudent choice for the individual. The occasions when the risk should be considered as well as the benefit will be discussed in chapter 9.

7
Power from the Atom

THE NEED FOR ENERGY

To assess the need and place for nuclear reactors in the generation of power, it is essential to review briefly the whole question of energy needs in our society. It is a subject which arouses mixed emotions, and so it is difficult to achieve a sense of perspective.

As a society becomes more industrialized, the demand for power increases alarmingly. When men and women lived a rural life and farmed the land, they asked only for food and warmth, and their energy needs were modest indeed. The energy they used came directly from the sun. The sun's rays ripened the crops, the process of photosynthesis using but a tiny fraction of the prodigal amounts of energy from the sun which fall daily upon the earth. Wood was burned in domestic fires, but it represented but a small proportion of surplus trees. Life was hard and sometimes cruel in those days, but man lived in a state of ecological balance. He used a little of the abundance of energy from the sun with which the earth is continually bathed, and did not deplete the stores of natural resources.

These days we are greedy for power. We expect to be warmed in winter and cooled in summer. We expect to travel whenever the need arises or the fancy takes us; by car to work, by high-speed jet to exotic vacations or distant business opportunities. The crowded air terminals and busy parkways bear testimony to the power used for these purposes; far more insidious is the vast amount of energy burned up in the manufacture of the material things which surround us. Iron ore is smelted to produce sheet steel for cars, refrigerators and washing machines. The extraction of aluminum is particularly costly in electricity, and much of it is used to make disposable beer and soft drink cans which seem to litter the entire countryside.

The human urge to acquire things appears to be insatiable. Each generation expects to be blessed with more material things and to enjoy a higher standard of living than the last. Like Oliver Twist before the Beadle, we each say in turn, "Please sir, I want some more." This characteristic of human nature was responsible for the staggering increase in power used in Western countries this century. As a rough rule, the demand for power in the United States doubled every 20 years or so until the oil crisis in 1974 brought an abrupt end to the upward spiral.

It may well be that we have now reached a time when our society will accept a leveling out in the standard of living, a break in the upward spiral of increased material wealth in the interest of preserving the environment in which we must all live, work and play. But whatever may happen about a future increase in the need for power, it is quite certain that the demand will not decrease, and it will be a problem even to maintain the present level of supply. It is in this context, therefore, that a brief review must be made of the generation of power from the various natural resources that are available to us on the earth.

ENERGY FROM FOSSIL FUELS

Almost all of the energy available on earth comes from the sun. The rays of the sun which continously bathe the earth represent a staggering amount of energy, but it is difficult to harness this energy and use it directly as we wish. A tiny fraction of the sun's energy (0.2 percent) is captured by plants in the process of photosynthesis, by which simple inorganic substances such as water and carbon dioxide are converted into plant tissue. The plants in turn are the food of animals, and either or both represent food for man.

On average, the rate of decay of dead plants and living things is approximately equal to the production through photosynthesis. However, during some geological periods, a minute fraction of this material was trapped in

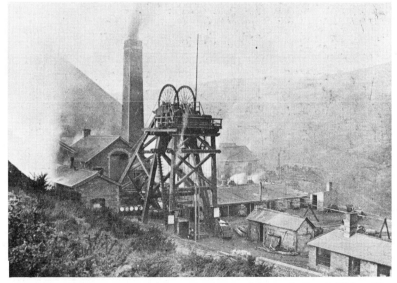

FIGURE 7.1. Penrhiw Colliery, the last wooden head frame in South Wales. (Courtesy of Dr. W.G. Thomas and The National Museum of Wales.)

peat bogs, where a lack of oxygen prevented complete decay. This quirk of geological fate accounts for the earth's store of so-called fossil fuels—oil, coal, natural gas, peat, etc.

Human beings have been remarkably inventive in gaining access to extra energy sources. The sequence of events went something like this: Domestication of animals as beasts of burden was the first step in using more energy than the muscle power of the individual. This was followed by the use of wind to propel sailing ships and waterwheels to grind corn. The real breakthrough in energy utilization came some 900 years ago when people on the northeast coast of England, near Newcastle-upon-Tyne, found that the black rocks along the seashore, which they termed "sea-coal," would burn.

The mining of coal as a continuous enterprise soon spread to all the coal fields of Great Britain and Western Europe, and it was this source of energy which enabled these nations to develop as industrial powers. Coal fired the boilers of the iron battleships of the Royal Navy. Coal was the fuel used in the smelting furnaces of the Ruhr Valley. The United States was a late starter, but has long since overtaken everyone else except Russia, who is the world's leading producer. By 1970, coal was being mined at a rate of over 3,000 million tons per year. This represents about one-third of the total energy extracted and used from fossil fuels, the remainder of course coming from oil and natural gas. Despite these staggering tonnages, it has been estimated that

FIGURE 7.2. The village of Nantyglo in South Wales during the Industrial Revolution, showing the devastation and pollution caused by the early coal and iron industries. (Courtesy of Dr. Morgan Rees and The National Museum of Wales.)

FIGURE 7.3. One of the giant mobile rigs used by Esso/Shell to search for oil in the North Sea. It has a displacement of 12,210 tons and can operate in water depths to 600 feet. This particular rig went into service in 1967 and made an oil strike in the Auk field off the east coast of Scotland. (Courtesy of the EXXON Corporation.)

total coal consumed from the start of mining operations in the twelfth century up to 1970 is only about 2 percent of the total stocks known to exist on earth. In other words, there is still an ample supply; the earth's resources of coal will not be exhausted for hundreds of years.

The picture concerning oil is quite different. Production of crude oil in the United States did not begin until 1859, but it has grown at an astonishing rate, which has only begun to slow in the past few years. While coal deposits cover enormous areas, oil and gas deposits are usually found in pockets of limited volume, trapped in spaces in the rocks. Despite an intensive search, new deposits of oil are becoming increasingly more difficult and expensive to find.

In the United States there are a vast number of oil wells, more than half a million, most of which are small producers. On the average, each produces only 20 barrels a day. Arabia has a smaller number of wells, but each on average produces 20,000 barrels of oil each day. More than half of the known oil deposits on earth are in the Middle East, while the principal consumers are the industrial nations of Europe, Japan and the United States. While it may be possible for the United States to become largely self-sufficient, and use only oil produced within her borders, Europe can never expect to satisfy more than half its requirements from its own off-shore deposits, and Japan must forever depend almost entirely on imports. This geographical distribution of oil makes it a vital bargaining point in international affairs.

It has been projected that if the use of oil continues to expand at its present rate, production will reach a peak by the year 2000, and subsequently decline to essentially zero by the year 2100. These projections must, of course, be taken with the proverbial pinch of salt. Prophets of doom are for the most part ignored, and history has usually proved the optimists to be right in the end. This is inevitable since circumstances change so rapidly that prophecy is at best a risky business; who knows, for example, how much off-shore oil is waiting to be discovered tomorrow?

Nevertheless, while not taking the details of the projections too seriously, it is evident that the supply of fossil fuels is strictly limited. Once they are burned, they are gone forever, and nature is not replacing them at a fraction of the rate at which we are now using them up. This is not a problem for tomorrow, or even for next year. But the long-range planners must look ahead beyond their own lifetime to that of their children and grandchildren. On this time scale, coal and oil reserves are fast running out, and alternative sources of energy must be urgently sought.

Compared with the million years or so that man has been on earth, the availability of fossil fuels has been a brief and transitory interlude—from a few hundred years in the past to a few hundred years in the future. At the same time it cannot be denied that the impact made by the burning of oil and coal on the ecology of our planet, and the way in which it has revolutionized our life style, has been unequaled so far by any other event in the history of the world.

SOLAR POWER

Incredibly large amounts of energy reach the earth every day in the form of heat and light from the sun. This solar energy maintains the earth's moderate climate, provides the energy requirements for plant and animal kingdoms by photosynthesis, and keeps the water cycle in operation by means of evaporation and rainfall.

If only a small fraction of this solar energy could be trapped and used directly, man's energy problems would be solved. There are two principal difficulties involved in harvesting this energy and converting it into conventional electrical power. First, the energy flux is intermittent; there is the regular cycle of night and day, and superimposed upon this there are variations due to weather conditions and in particular the extent of cloud cover. Second, the amounts of energy involved are worth considering only when they are integrated over large areas. Regions close to the equator are obviously more favorable because of the greater intensity of sunlight.

The most suitable land areas, fortunately, turn out to be those that are sparsely covered with vegetation and almost unpopulated; included would be part of the southwestern United States, much of northern Mexico, a large coastal belt in Peru and Chile, the Sahara Desert in northern Africa, the Red Sea area and Persian Gulf and West Pakistan. In the case of the United States, Arizona would be an obvious contender. The problem is the cost. In order to equal the electricity produced by a large conventional power station, 20 square miles would have to be covered with solar collector plates to harvest the sun's rays. With present-day technology, the cost would be prohibitive, but this may change in the future due to technological advances. The attraction of solar power is that it would utilize dry and barren areas of land and is essentially pollution free, although some critics point out that we do not know the consequences of molds that may grow under 20 square miles of collector plates and cause a new problem. It has been estimated, for example, that if a fifth of the total land area of Arizona was devoted to the conversion of solar energy into electrical power, it might be possible to satisfy the total needs of the United States. Estimates such as this are easy to make, but are of little use. In practice, solar energy will make a small, but steadily increasing, contribution to our total energy needs. Over a prolonged period of time, the design of new houses can incorporate solar systems to provide space heating and domestic hot water; at the present time it is not an economical proposition to add solar systems to existing houses, though this would change if costs of fossil fuels rise dramatically. It is of interest to note here that the risk to life and limb of the initial installation, followed by periodic cleaning and maintenance of solar panels on the roofs of houses to provide hot water, is far greater than the risks from nuclear power plants to produce electricity to heat water! A major utilization of the use of solar energy cannot come until cheap and reliable photo-voltaic cells are available to directly convert sunlight to electricity. This awaits a technological breakthrough which may take a number of years. To put things into perspective, the forecast of what solar energy might contribute to the United States energy supply by the year 2000 has just been doubled—from 1.5 percent to 3 percent!

HYDROELECTRIC POWER

The generation of electricity by the flow of rivers returning to the sea has already been exploited to a considerable extent, principally in the highly industrialized areas of North America, Western Europe and Japan. The largest hydroelectric complex in the world is being built in the Quebec province of Canada. In the vicinity of James Bay they are changing the landscape: reversing rivers, damming valleys and flooding an area the size of the British Isles. Since Northern Quebec's hard, rocky soil doesn't yield silt to fill the reservoirs prematurely, and its frigid climate is not subject to draught, it produces the cheapest electricity in North America; much of the output is sold across the border to New York State. Most observers think that the Quebec hydro scheme has been hopelessly bungled from the outset and has cost twice as much as it should. But, nevertheless, it produces cheap power—which will only appear cheaper still in the future as inflation escalates the price of other forms of energy. The underdeveloped regions of Africa, South America and Southwest Asia have the largest potential water power capacities, which will almost certainly be developed sometime in the future. It has been estimated that if all of the potential water power in the world were fully exploited, it would equal approximately the world's present rate of industrial power use.

At first sight this appears to be an ideal source of power, inasmuch as it is pollution free and supposedly inexhaustible. This may not be true, however, since most hydroelectric schemes involve the creation of reservoirs by the damming of streams. Within a few hundred years, these reservoirs fill up with sediment, and unless a technical solution can be found for this problem, water power may actually be comparatively short lived.

An alternative way to generate electrical power from water would be to utilize the motion of the tides, instead of the unidirectional flow of rivers. Such schemes involve trapping sea water at high tide and using it to generate electricity as it later flows back to the sea. The world's first large-scale tidal power plant, which began operations in 1966, is on the Rance estuary on the English Channel coast of France. There are a limited number of places in the world where a combination of high tide and a suitably shaped bay or estuary would make similar projects feasible and practicable.

Hydroelectric schemes are a most useful way to produce electrical power. Every suitable site will eventually be exploited, but the total capability is so restricted that it will never make more than a limited contribution to the solution of the energy crisis.

GEOTHERMAL POWER

Any visitor to Yellowstone National Park must have contemplated how convenient it would be to connect his own central heating system to one of the natural hot springs and enjoy an abundance of free heat and hot water forever. Suitably harnessed, "Old Faithful" would solve the energy crisis for a small town, now and for a long time to come!

Geothermal power has been harnessed in a few volcanic areas of the world, where natural steam is available at or close to the surface. Examples are the Tuscany region of Italy, where geothermal power has been produced since 1904, Wairakei in New Zealand, Iceland, Indonesia, the Philippines, and the geysers of northern California. Small plants also came into operation recently in the USSR, Mexico and Japan. The amounts of power involved in all of these cases is trivial compared with national needs, although they represent an invaluable resource to the local area.

Looking to the future, ambitious plans have been proposed to tap the virtually inexhaustible heat stores in the center of the earth. The occasional eruption of a volcano and the widely scattered hot springs on earth give us a hint of the prodigious amounts of energy locked within our planet, waiting to be exploited. To harvest this heat source would require drilling a hole many miles deep through the crust of the earth, which would involve the development of a completely new technology. It also raises the question of what would happen if the earth's crust were artificially pierced in this way. Would molten lava spew from the hole to constitute the first man-made volcano? Would all the molten material leak out of the planet and change its magnetic properties?

Geothermal projects in many parts of the world are harnessing steam and hot water from beneath the earth's surface to generate electricity; however, serious problems threaten geothermal prospects everywhere. The principal problem is its cost. Most experts agree that projects based on "dry steam" can succeed. But most of the world's geothermal resources are found, not as naturally distilled steam, but as hot ground water containing corrosive salts and other solids that are difficult and expensive to handle. Meanwhile, environmentalists complain of spoiled natural vistas, escalated noise levels and the smell of rotten eggs from the hydrogen sulfide emitted from these deep underground geothermal sources. For example, the once-impressive array of 15 steaming geysers in Rotovoa, New Zealand has shrunk steadily to five—a disappointment to tourists and an ominous sign that extensive development can deplete this natural resource.

To sum up, geothermal power is at present an important resource in a few local areas, but as a solution to the national or worldwide energy crisis, it is no more than a fascinating idea which needs to be checked out.

NUCLEAR POWER

The source of stored energy most recently exploited is that trapped in the atomic nuclei of some of the earth's natural constituents. Nuclear energy is released by two opposite processes.

1. If the nucleus of a large atom, such as uranium, is torn apart to form two smaller atoms, surplus energy is released, mostly in the form of heat.

This process, whereby a heavy atom is broken into fragments, is known as *fission*. It is the basis of the original atomic bomb, and it is the principle upon which all nuclear reactors for physics research and power generation are based.

2. Alternatively, if the nuclei of two light atoms, such as hydrogen, are joined together, or fused, to form one larger atom, energy is again released. This process is known as *fusion*. Under high pressure and intense temperature, where there were two atoms, there is now one. This event is accompanied by a loss of mass and a resultant release of a large quantity of energy.

Fusion has a prototype, and one we can all see in action. The sun, which our forebears thought to be a ball of burning gas, is in fact a mass of nuclei fusing together at extremely high temperatures. It is also the principle of the hydrogen bomb, which releases a prodigious amount of energy. Every major nation in the world, as well as a consortium of countries in the European Economic Community, has a large and expensive program attempting to harness the fusion process in a controlled device to generate power. Despite the effort and money spent, no one is close to a practical system for producing fusion power. It is an attractive possibility and a big hope for the future because its basic fuel is essentially inexhaustible and the principal radioactive waste is tritium, far less of a problem than the by-products of the fission process. However, it is not certain that fusion will ever be practical; it certainly cannot be exploited now or in the near future, and so it will not be discussed further in this chapter.

Fission

Isotopes of uranium and thorium are the only fissionable materials found in nature, and of these, uranium-235 is by far the most useful. When uranium is harvested from its ore, it consists mostly of uranium-238; only one of each 141 atoms is uranium-235, which is the primary fuel source used in present-day nuclear power plants. When an atom of uranium-235 is struck by a stray neutron, the neutron may be absorbed and cause the atom to undergo fission, that is, to break up into two roughly equal parts, comprising two complementary atoms somewhere in the midrange of the atomic scale. When fission occurs, the enormous energy that holds the nuclei together is released. The total mass of the flying fission fragments is less than the mass of the original nucleus. The mass that is lost is changed into energy, as Einstein had predicted in his famous equation:

$$E = mc^2$$

where E is the energy released, m is the mass destroyed and c is velocity of light.

Although the mass lost is tiny, the energy released is enormous and much of it is available as heat. The fission of one uranium nucleus produces a million times more energy than the explosion of one molecule of TNT. The energy released in the most powerful chemical reactions known to man are trivial

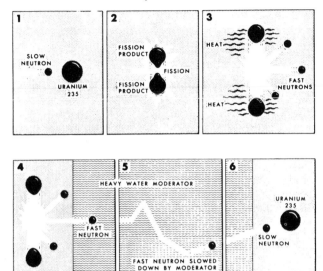

FIGURE 7.4. The nuclear fission process. Slow neutron (1) strikes nucleus of ura-
nium-235 atom and splits ("fissions") it (2) into fission products which fly apart
(3) creating heat. Neutrons given off at the same time are slowed down as they travel
through moderator (5). (Diagram from Atomic Energy of Canada Limited.)

compared to the energy of fission. The fission fragments into which the
uranium atom is split up include a large number of possible elements; in fact,
over 100 combinations are possible. Most of these fission fragments are
radioactive, some of the most potentially hazardous being radioactive iodine,
cesium and strontium. The lifetimes of these fragments also vary enormously; some decay in a few seconds, minutes or hours, while others last for
many thousands of years.

After the fission of each uranium atom there is always a surplus of
neutrons, usually two or three. There is a possibility that these neutrons may
hit other atoms of uranium-235 and split them into fission fragments, too,
plus, of course, still more neutrons. If the three original neutrons ejected
from one fission cause three more fissions, and the nine neutrons released
each cause a further fission, 27 neutrons will then be available. It is at once
obvious that the number of neutrons and the consequent number of possible
fission events will increase at a very rapid rate. This is called a chain reaction,
and builds up quickly within the uranium fuel element./

THE ANATOMY OF A NUCLEAR POWER PLANT

In conventional electricity generating plants, fired by coal or natural gas,
the fossil fuel is burned in a firebox, and the flaming heat generates steam in
the boiler. This steam, the historic propellant of the industrial age, roars into

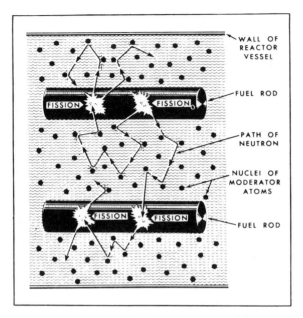

FIGURE 7.5. Chain reaction. Neutrons given off by splitting ("fissioning") uranium-235 atoms are slowed down (their speed is "moderated") as they bounce against the nuclei of the moderator, which may be heavy water, ordinary water or graphite. Slowed sufficiently, the neutrons will split other uranium-235 atoms and thus maintain a chain reaction. (Diagram by Atomic Energy of Canada Limited.)

a huge turbogenerator at a high pressure, sometimes reaching 2,700 pounds per square inch, and at a searing temperature of up to 1,000 degrees. The steam spins a huge turbine, whose shaft is coupled to a mammoth generator which produces electricity. These new steam electric plants generate more than one million kilowatts of power. The fossil fuel appetite of such a plant is prodigious; if coal is used, more than 400 tons is required each hour.

A nuclear plant "burns" the flameless fuel uranium; heat is released as atoms split in a controlled chain reaction. Since there is no combustion, there are no exhaust gases and of course no sulphur or carbon dioxide pollution. The nuclear firebox consists of a compact core with a volume smaller than an average living room. It contains a year's supply of nuclear fuel—more than 100 tons of thimble-size uranium oxide pellets. About 10 million of these tiny pellets are neatly arranged in 12-foot tubes, or fuel rods, encapsulated to prevent leakage of radioactivity. The nuclear fuel used in modern power plants contains only a few percent of uranium-235, compared with about 90 percent enrichment in weapon grade material, and so there is no possibility that a fission reactor will explode like a bomb. However, it is still potent stuff—a single half-ounce pellet releases the same amount of energy as 160 gallons of oil.

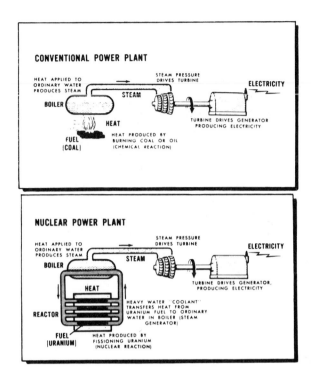

FIGURE 7.6. Main difference between a conventional steam power plant (upper panel) and a nuclear plant lies in the source of heat used to make the steam that drives the turbines. In the reactor (lower panel) the primary coolant (water or heavy water) flows over hot uranium rods and becomes hot itself. It is then pumped through a boiler (heat exchanger) where it gives up its heat to ordinary water that is converted into steam. (Diagram from Atomic Energy of Canada Limited.)

To start and sustain a chain reaction, the fuel rods must be held in a core of some suitable substance, preferably a light element, the purpose of which is to slow down, or to "moderate" the neutrons produced by the fission of uranium-235. When these neutrons are first "born" during the break-up of a uranium atom, they move at high speed, but oddly enough they are more effective at splitting other uranium atoms if they are first slowed down by bumping into other atoms of a light element in the core. There are many possibilities for the material of the core or moderator. The three most common are graphite (i.e., carbon), ordinary water or "heavy" water, i.e., water in which the hydrogen is replaced by deuterium, a heavier isotope of hydrogen.

A discussion of the core of a nuclear reactor may seem to be an unduly technical detail and out of place in a book of this sort, but it turns out to be a vitally important factor in understanding the design of commercial nuclear power plants.

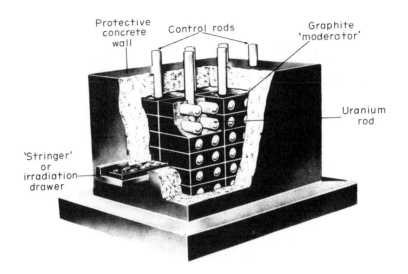

FIGURE 7.7. Diagram to illustrate the principle of a nuclear reactor. (Courtesy of Drs. Meredith and Massey and John Wright & Sons Ltd.)

Safety Mechanisms

The key elements in reactor safety are the three Cs—Control, Cooling and Containment. Reactors must be designed, constructed, operated and inspected so that the probability of failure of any of these key elements is vanishingly small, because the consequences of a failure would be the release of enormous amounts of radioactivity into the environment. Design is based on redundancy, with multiple parallel systems so that if one breaks down, another can take over. This is particularly important for the cooling systems.

Control Rods

The reactor is controlled by inserting into the core rods a material such as boron or cadmium, which tends to absorb, or swallow up, neutrons. With programmed withdrawal of these rods, the chain reaction develops; by a continual adjustment of these rods, pushing them in or out of the core, the level of operation of the reactor can be kept at the desired level.

Removing Heat from the Reactor

Nuclear reactors are used for many different purposes. In a physics research laboratory, the most interesting property of a reactor is the availability of a high flux of neutrons within it. These may be used to perform

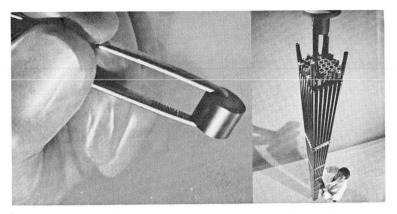

FIGURE 7.8. Pellets of enriched uranium (left) and fuel assemblies (right) as used in a United States light water reactor. The pellets are inserted into rods, and the rods are grouped to form assemblies which make up the core of a nuclear reactor. (Courtesy of EXXON.)

experiments in nuclear physics, or may be used to bombard targets and produce radioactive isotopes for research, medical and industrial uses. In this context the tremendous energy that is dissipated in the form of heat is a nuisance and must be removed by some kind of cooling system. On the other hand, when reactors are used to generate electricity it is the heat produced that is of major importance. The radioactive features of the reactors are an unfortunate characteristic, and indeed a nuisance which must be carefully contained and shielded.

In a power reactor the prodigious amounts of heat produced in the core must be removed continually as steam and transferred to the turbines, which turn the generators and produce the electrical power. This is accomplished in one of two ways.

In reactors having a core or moderator made of graphite (i.e., carbon), the heat is flushed away and the core cooled by flowing a gas through it. On the other hand, in reactors using light or heavy water as a moderator, the water itself can be circulated out of the core to remove the heat. Whether water or gas is used for cooling, it is imperative that the method of heat removal should be both effective and continuous, otherwise the core and fuel containers could melt and release vast quantities of radioactivity. It is equally important that the control system should be sensitive and operate efficiently so that the rate of heat production does not, even briefly, exceed the capacity of the cooling system. The reactor can be shut down rapidly in the event of a suspected malfunction, but even then continued cooling is essential because although no heat is produced by fission, heat continues to be generated by the decay of residual radioactivity in the fuel. Immediately after shut-down, this amounts to about 5 percent of the heat at full power, so continuous cooling of the fuel is still an absolute necessity.

Shielding

Since fresh fuel is only mildly radioactive it can be handled without shielding. After the reactor has operated for some time, the fissionable material has been partly used up, and the activity will have increased about 10 million times due to the creation of the radioactive fission products in the fuel. It is for this reason—to shield the products of fission—that so much shielding is required around the core and why it is so vital that the containment vessel should not be breached, otherwise this radioactivity could escape.

REACTOR TYPES USED FOR POWER PRODUCTION

The commercial production of electrical power from the atom is already an accomplished fact. It is not always appreciated that there are such basic differences in the types of power reactors that have been developed and used in different countries.

(i) *British reactors.* Because of the urgent shortage of fossil fuels, and a great dependence upon imported oil, Britain was the first country to make a significant commitment to nuclear power. As early as 1955, when the riches of North Sea oil were not known, they opted to use reactors with a graphite core, cooled by a gas (carbon dioxide). The hot gas from the reactor is used to produce steam and generate electricity in huge turbo-generators. The older Magnox reactors, named after the magnesium alloy fuel containers used in them, contain about 50 tons of uranium metal of natural composition; in the early days, the fact that the fuel did not need to be enriched with uranium-235 was considered to be an attractive feature of their design. The newer, advanced, gas-cooled reactors contain about 100 tons of slightly enriched uranium oxide. In both reactor types, massive concrete shielding is provided because of the intense radiation eventually emitted by their use. Both the reactor and heat exchanger are housed in a containment building. Unfortunately the efficiency of the early Magnox reactor is poor because high temperatures and pressures cannot be used since carbon dioxide turns out to be highly corrosive under these conditions.

The Berkeley nuclear power station situated on the Severn estuary in England was completed in 1962 and represents the first fully commercial use of the atom to generate electricity. A number of similar reactors, of increasing size, have been commissioned over the years and now produce a significant proportion of the electricity consumed in Britain, but no buyers have been found for them in other countries. The more modern British designs are extremely attractive from many points of view, but they lost the race for export orders because they could not muster the economic strength to achieve the level of engineering perfection necessary in world competition. Indeed, future nuclear power stations in Britain are likely to be equipped with light water reactors similar to those used in the United States.

(ii) *U.S. reactors.* In the 1950s there appeared to be an abundance of fossil fuels in the United States, particularly oil, and as a consequence there was no urgent need to make an early commitment to nuclear power. Many different reactor designs were studied, but by 1957 the United States Atomic Energy Commission had focused attention on two types of reactors, both of which

FIGURE 7.9. Fuel element bundle from the Canadian CANDU heavy water reactor. (Courtesy of Atomic Energy of Canada Limited.)

used light water as a moderator (i.e., ordinary water). The final development of these reactors was left to private industry, and several models are now available commercially. Many are in use in the United States, and some have been sold to other countries. The two principal competing types will now be described briefly.

The boiling water reactor (BWR) is shown schematically in Figure 7.10. In this system the coolant, which is ordinary water, is allowed to boil in the reactor vessel. The steam is collected in the dome of the pressure vessel, and goes directly through pipes to the turbine, which drives a generator to produce electrical power. It is important to recognize that in the BWR, any radioactivity in the coolant water can be carried over with the steam into the turbine. There are elaborate filtering systems to trap the radioactivity, but some inevitably escapes and the emissions from a BWR are higher than for other types of reactors. The fuel consists of uranium oxide (UO_2) pellets in which the uranium is enriched to contain about 2 to 3 percent uranium-235. The pellets are sealed into fuel rods or pins, typically 12 feet long and containing about 500 lb of uranium oxide. The core of a 1,000 megawatt reactor would contain about 760 such elements totalling about 186 tons of uranium oxide.

The alternative design used in the United States is the pressurized water reactor (PWR), shown schematically in Figure 7.11. In this type of reactor, the water in the core is under high pressure to prevent boiling. The heat from the water is transferred to a secondary loop containing water at low pressure, which boils to form steam and drives the turbines to make electricity. The significant point is that the steam produced should not contain any radioactive products from the reactor core. In practice, there are tiny leaks between

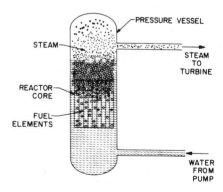

FIGURE 7.10. Schematic diagram of a steam source provided by a Boiling Water Reactor (BWR), one of the two types of light water reactor manufactured in the USA. The reactor core causes the water to boil, and steam to form at the top of the pressure vessel. (Courtesy of Brookhaven National Laboratory.)

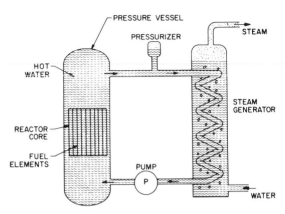

FIGURE 7.11. Schematic diagram of the Pressurized Water Reactor (PWR), one of the two types of light water reactors manufactured in the USA. The steam is formed in a secondary loop and is isolated from the water which surrounds the reactor core. (Courtesy of Brookhaven National Laboratory.)

primary and secondary systems, but the exhaust steam from a PWR contains less radioactivity than that from a BWR. The fuel is similar, but not identical, to that used in the BWR.

These two reactor types are manufactured by rival companies which compete for orders from the utilities that generate electricity. A very important factor in the successful design of either the boiling water or high-pressure water reactor is the huge steel pressure vessel that is required to house the reactor core. These tanks, which are approximately 30 feet in diameter and 70 feet long, are fabricated of steel 8 inches thick. They must withstand a pressure of several thousand pounds per square inch, and high temperatures as well. A specialized industry has grown up to manufacture reactor pressure vessels, which are among the largest objects ever fabricated by man. Stringent quality control, x-ray examination and frequent testing is required during manufacture and in subsequent use, because the safety of the reactor depends vitally on the integrity of this steel container. It has been estimated that if adequate care is taken in both manufacture and maintenance of this steel vessel, the chance of an accident should be reduced to less than one in a million years of operation.

(iii) *Canadian reactors*. In Canada, which has a vigorous nuclear industry, yet another type of reactor has been developed for commercial use. In this reactor, the moderator is heavy water (deuterium oxide) which also serves to cool the reactor. A heat exchanger is used to transfer the heat from this very valuable heavy water in the primary loop to ordinary water in the secondary loop of the heat exchanger, which powers the turbines to generate electricity.

(iv) *High temperature gas-cooled reactors.* A number of countries, including Britain, France, and Germany, as well as the United States, are developing high temperature reactors, based on a graphite core with an inert gas such as helium to cool the core and transfer the heat to the turbines. This design has several significant advantages.

First, a higher temperature results in improved efficiency, and 900°C can be maintained because an inert gas such as helium is used instead of the carbon dioxide in the old British graphite reactors. It should also be compared with a maximum of 300°C in the water-cooled American or Canadian reactors.

Second, a gas cooled reactor is inherently much safer than one that is water cooled—the likelihood of a serious accident or leaks resulting from a breakdown in the cooling system is much less than in water cooled devices.

Third, the fuel element design, at least in the German version, allows fuel to be replaced continuously as it burns up without the necessity of shutting down the reactor for refueling.

While clearly superior in principle, the high temperature gas cooled reactor is not at present commercially competitive because of the capital costs involved.

JAPANESE NUCLEAR TECHNOLOGY

Japan—the only country on earth to have suffered the devastation of an atom bomb attack—has long since developed its own nuclear technology. In a country with the horrific legacy of Hiroshima and Nagasaki, there is a refreshingly sharp distinction between nuclear weapons and reactors for power production. Consequently, while nuclear weapon protests regularly draw large crowds, the ambitious nuclear power program enjoys wide support. Japan lacks domestic oil and coal, and at the time of the 1974 oil embargo imported 80 percent of its oil from the politically insecure Middle East. The Japanese government quickly devised a plan to produce 17 percent of the country's power by nuclear reactors by 1985—a goal already achieved substantially ahead of schedule. Japan's nuclear capacity is expected to triple by 1990 with the addition of 17 new reactors, by which time close to half of Japan's electricity is scheduled to come from nuclear sources, second only to France in its reliance on nuclear energy.

No plants have been cancelled because costly political and legal delays rarely occur in Japan. This is a reflection of the very different social and political climate in Japan. The government, aided by academics and industrial spokesmen, have "educated" the public to regard nuclear power as vital for Japan's prosperity and nuclear plants as an employment opportunity.

THE BREEDER REACTOR

When uranium ore is mined, it consists largely of uranium-238. Only one part in 140 is uranium-235, the fissionable fuel used in a conventional reactor. Consequently, only a small fraction of the energy contained in the world supply of uranium can be released and used with present-day commercial reactors. Fission can occur in uranium-238 in a specially designed *breeder reactor.* This has no moderator to slow down the neutrons, since fission occurs by the capture of high energy or fast neutrons by uranium-238. This process is known as *fast fission.* The reactor is cooled with liquid sodium and the heat can be extracted and used to produce power in the usual way. At the same time, some of the neutrons released by the fission of uranium-238 are captured by other atoms of the same element to form plutonium-239; this material is itself fissionable and can be used as a fuel in conventional reactors. Thus, by using breeder reactors which utilize uranium-238, it is possible to burn up and produce energy from essentially all of the uranium in the natural ore while at the same time producing fuel (plutonium-239) for conventional reactors. In short, if breeder reactors are used in parallel with conventional reactors, then the amount of energy that can be extracted from the world supply of uranium is increased about one hundred fold. At first sight it would seem that the design and development of the breeder reactor would be a high priority item, but there are political considerations involved. For the breeder reactor to be usable, plutonium must be extracted from the used spent fuel elements to be used then in conventional reactors, and this is perceived to be dangerous since plutonium is also useful for the manufacture of nuclear weapons. Consequently, the development of a breeder reactor is held up by the desire to limit the proliferation of nuclear weapons. Of course, uranium-235, the commonly used fuel of commercial reactors, may also be used to make an atom bomb, but it is much more difficult to separate from uranium-238 in sufficient purity for a weapon.

COMPARING REACTOR TYPES

The whole subject of reactor design is of enormous complexity and involves political considerations and national pride as well as questions of science and technology. However, there are a few simple factors which are easy to understand and which have important implications. The carbon dioxide-cooled, graphite-cored reactors have been used successfully in Britain for many years and have already been worth their development. However, they are now obsolete and are no serious rival for the more modern reactors developed in the United States and Canada, the two principal countries that compete for overseas sales of nuclear power plants.

The advocates of the light water reactors, commonly used in the United States, point out that they are the most attractive from an economic point of view. They certainly cost less to build, which is an important consideration at a time when investment capital is at a premium. If the production of cheap electricity is the primary concern, then they would always be favored over the heavy water moderated reactors built in Canada. However, there are other factors to be considered. Since the United States reactors will not operate

1 England   **2 England/Germany**

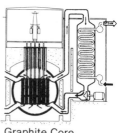

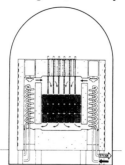

Graphite Core
CO_2 Cooled

5

High Temperature
Graphite Core
Helium Cooled

3 United States

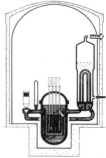

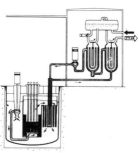

Fast Breeder Reactor

4 Canada

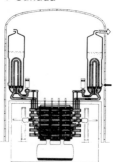

Light Water Cooled
Enriched Uranium

Heavy Water
Natural Uranium

FIGURE 7.12. Illustrating some of the variety of nuclear reactors used, planned or contemplated for power generation. *1*. The earliest and simplest power reactors have a graphite core, are gas cooled with carbon dioxide and use natural uranium. They have been used for commercial power generation in Great Britain for over a quarter of a century. *2*. Several countries have research and development programs to improve the efficiency of inherently safe gas cooled reactors by raising the temperature and using an inert gas for cooling. *3*. The U.S. reactors were designed originally to power submarines and aircraft carriers. They are very compact, cooled with ordinary water, but need enriched uranium. *4*. Canadian reactors are very large, use natural uranium and have a core of heavy water. *5*. If full use is to be made of nature's supply of uranium, fast breeder reactors must be developed to convert uranium-238 into fissionable material. This is a hot political item; the fissile material made in a breeder reactor can be used for bombs as well as power production. The Breeder program is barely kept alive in the U.S., no one knows for sure how many other countries are working on this secretly.

FIGURE 7.13. The Berkeley nuclear power station of the British Central Electricity Generating Board, situated on the eastern bank of the Severn Estuary, started in 1957. This is one of the earliest commercial nuclear power stations. Note the two reactor buildings, housing MAGNOX graphite cored gas cooled reactors, with eight boilers grouped around each. (Courtesy of the Central Electricity Generating Board of the U.K.)

with naturally occurring uranium, their fuel must be enriched artificially with uranium-235. Consequently, any country purchasing such a reactor must also have some long-term treaty to be supplied with enriched uranium, which means that the United States keeps control over the use of the reactor for all times in the future. In fact, this situation is unlikely to last long, since the USSR, as well as other countries with a nuclear industry, is now prepared to sell enriched fuel.

The Canadian reactor was designed specifically for commercial power production and is moderated with heavy water. Consequently, it can operate with naturally occurring uranium, and any country which buys a reactor of this type is not dependent on the United States for a continuing supply of enriched fuel. To be sure, they cost more to build in the first place, partly because of the cost of heavy water and partly because they are six to seven times as big as the compact American light water reactors, which are in

FIGURE 7.14. Aerial view of a nuclear power station housing a United States light water reactor. Note the compact size for a plant producing 626 megawatts of power. It is located on the shore of Lake Ontario in upper New York State and began operating in 1969. (Courtesy of Niagara Mohawk Power Corporation.)

FIGURE 7.15. The showpiece of the Canadian nuclear power program is Pickering Generating Station, just east of Toronto, which was completed in 1973. It has four heavy water reactors with a total capacity of 2000 megawatts. (Courtesy of Atomic Energy of Canada Limited.)

essence modified versions of the atom plants originally conceived to power submarines and aircraft carriers. However, their makers claim that over a five-year period they break even in operating costs compared with the U.S. reactors because the fuel is cheaper and is burned up more completely; about twice as much electricity is produced for each pound of uranium "burned". The heavy water reactor also offers the opportunity to produce plutonium more efficiently, which is a material that can be used to make an atom bomb or recycled back for fuel enrichment. The Canadian reactor, therefore, is the obvious choice of any nation which is not simply interested in producing electricity at minimum cost, but also seeks to join the nuclear club. This, of course, was the route by which India made its first atom bomb in the early 1970s, much to the embarrassment of the Canadian government, who had generously supplied several CANDU reactors as part of their aid program!

An unfortunate feature of a CANDU reactor is that it produces much more radioactive tritium—about 1,000 times as much as a light water reactor. The tritium is produced in the heavy water moderator; a typical reactor may contain about a kilogram of tritium, which is a very small proportion of the thousands of gallons of heavy water, and initially it was thought that it could be left in place and ignored. However, in practice it poses an additional hazard to workers around the reactor, who quickly build up a body burden of tritium to the point where their time spent in certain areas may be restricted.

Consequently, it is necessary to use a separator to constantly remove the tritium from the heavy water of the reactor. This is an additional complication and expense on CANDU reactors, but it is viewed with considerable interest in the fusion program because if this ever proves to be a practical way to produce electricity, then the release of radioactive tritium will be the major hazard. A fusion plant would contain at least as much tritium as a CANDU reactor, and so the separation and handling of large amounts of tritium, being developed by Canadian scientists, is of critical interest to fusion researchers all over the world.

A NATURAL REACTOR

Nuclear fission reactors are usually associated with the twentieth century, and thought to be the product of man's inventive genius. In fact, boiling water reactors have occurred in nature in the remote past. Several examples were found recently by French engineers in a region of Africa that is now the Gabon Republic.

They first noticed a slight deficit in the expected content of uranium-235 in ore shipped from a mine near Gabon. The only plausible explanation was that some of the uranium-235 had been consumed as a fuel in a natural nuclear reaction. The idea was quickly confirmed by identifying the radioactive waste products in the ore, formed by fission. What must have happened is as follows:

Natural deposits of uranium ore were located in a rock structure with water flowing through it. The water acted as a moderator, slowing down neutrons emitted by spontaneous fission, and allowing a chain reaction to be initiated and sustained. Heat was generated and the inevitable radioactive waste products produced, some of which are, interestingly enough, still safely trapped in the rocks. In times of drought when the water supply would dry up, the reactor would stop—only to start again when the flow of water resumed. They were, in short, natural reactors, which operated at a power level of several kilowatts for at least 100,000 years and possibly for more than a million. It is estimated that the reactors were active about 1.7 billion years ago. Today, naturally occurring uranium will not sustain a chain reaction with ordinary water as a moderator because only one atom in every 140 is uranium-235; the remainder is uranium-238. Billions of years ago, the proportion of this fissile uranium isotope in natural ore was much larger, about 3 percent, which is comparable to the enriched fuel used today in commercial light water reactors. The proportion of uranium-235 has decayed over the years because its half-life is shorter than uranium-238. There are probably 50 to 100 other locations in the world where a combination of rock conditions and mineral deposits could have resulted in natural reactors. There is, truly, nothing new under the sun, and that includes nuclear reactors!

THE AVAILABILITY OF NUCLEAR FUEL

The success of nuclear fission as a replacement for fossil fuels in the generation of electricity depends mainly on an adequate supply of uranium in an economically extractable form. There are vast quantities of ores containing uranium and thorium in various parts of the world, and of the Western nations the United States has by far the richest deposits. The high-grade ores now being mined will provide an adequate supply for the present conventional reactors well into the next century at a cost which need not be excessive.

The long-term prospects of nuclear power depend on the development of the breeder reactor that uses uranium-238. Several small prototypes have already been built. With the advent of this device, it becomes an economic possibility to mine very low-grade uranium ores, which are available in vast quantities. For example, in the United States, a low-grade deposit, known as Chattanooga shale, crops out along the western edge of the Appalachian mountains in eastern Tennessee and underlies at mineable depths most of the areas of Tennessee, Kentucky, Ohio, Indiana and Illinois. It has been calculated that the uranium and thorium ores extracted from this shale over a single area 25 miles square could produce a total amount of power equal to the entire supply of fossil fuels in all of the United States. Thus, with the breeder reactor, the fission process could provide us with energy for at least a thousand years.

HAZARDS AND PROBLEMS
OF NUCLEAR POWER REACTORS

During the past two decades the opposition to nuclear power plants, particularly in the United States, has increased in both intensity and sophistication. The objections that have been raised, some real, others imaginary, have been the subject of heated debate and have seriously delayed the development of the nuclear power industry. In some countries, including the United States, the damage is probably irrevocable.

When it comes to public relations it has been an unequal struggle from the start. The objectors to nuclear power have made eloquent and emotional pleas, dramatized and exaggerated the dangers of reactors, and received wide coverage in the popular press. By contrast, the response of the establishment has been slow, voiced in cautious and ineloquent statements, and virtually ignored by the press. Not only so, but by the time the cumbersome machinery of a government bureaucracy is able to formulate a careful and reasoned answer to one question, the nimble-footed objectors are already off on a new tangent, forecasting disaster, death and destruction on entirely different grounds.

For this reason the following list may be out of date before this book is even published, but at the present time it includes the most justifiable fears and concerns that have been voiced concerning the proliferation of nuclear power reactors.

1. Thermal pollution
2. The mining of uranium
3. Routine releases of radioactivity
4. Processing and disposal of nuclear waste
5. Transportation of radioactive waste
6. Accidents
7. The homemade atom bomb
8. Nuclear proliferation

FIGURE 7.16. Aerial view of a nuclear power station in France; this nation has a highly developed and fast growing nuclear industry because of their reliance on imported oil.

1. *Thermal pollution.* The possible effects of waste heat discharged from electricity-generating stations into bodies of natural water is an unknown entity which must be viewed with caution. It surely does not represent a hazard to human life, but there is a potential damage to the environment. Waste heat is not confined to nuclear power stations, either. Conventional electricity-generating stations, burning oil or coal, must also disperse surplus heat, although in this case much of it is discharged directly to the atmosphere through the smokestack in the form of hot gases. Nuclear stations usually discharge most of their surplus heat into a river or lake.

If the nuclear plant is located on a large body of water, such as a very big lake or the ocean, then with reasonable precautions the impact upon the environment is minimized. However, construction of power plants on rivers or small lakes needs to be examined much more carefully, particularly in a highly industrialized area where several power plants may be built on the same waterway.

At the same time, water for condenser cooling for these large plants, particularly on once-through systems, is becoming more scarce as the energy requirements continue to expand. In England now, over 300 cooling towers are in operation for the control of thermal discharge, and in the USA it has been estimated that by the year 2000, over half of the available natural water runoff would be needed for steam-electric power plant cooling. A single 1,000-megawatt nuclear power plant, using once-through cooling, requires about $50m^3/sec$ of water for cooling, which is equivalent to the entire water consumption of a city of the size of Chicago.

Instead of regarding this heat as an unfortunate pollutant, attempts have been made to harness this energy in agriculture or aquaculture. Agricultural uses may include greenhouse heating, warm water irrigation, soil heating and frost protection of crops. In aquaculture, the use of heated effluents from power plants to maintain optimal temperature for growth and high yields of fish and seafood is only a recent development. The Japanese have led the way in demonstrating the benefits of waste heat; during the past 15 years, culture experiments with shrimp, eel and seabream have been carried out with thermal effluents from conventional power stations. Since 1971, a large culture program involving shrimp and fish has been in progress at Japan's first nuclear power station, the Tokai reactor of the Japan Atomic Power Company. Another successful use of waste heat is the commercial operation by the Long Island Oyster Farm in the USA, which utilizes thermal effluents from a commercial plant to enhance growth during the early stages of oyster culture. The normal growing period of four to six years has been reduced to two-and-one-half to three-and-one-quarter years by selective breeding, spawning larval growth and by seeding the oysters in the hatchery and then placing them in a warmed discharge lagoon of the plant for about six months.

2. *The mining of uranium*. Uranium has been found in many areas of the world, but the major sources are in Czechoslovakia, Australia, North America, and South Africa. The principal mining sites in the United States are in New Mexico and Wyoming. Like coal, uranium can be dug from underground mines or obtained from surface strip mining. A typical ore may contain only two parts in a thousand of uranium oxide so that large amounts of ore must be removed and treated in order to separate out the required uranium. These are called the "milling" sites. After the uranium has been separated, large quantities of residue or "tailings" must be disposed of. This is composed largely of crushed rock or sand, but it also contains tiny amounts of uranium, radium and other radioactive daughters. The principal hazard is that the radioactive gas, radon, is given off by the "tailings". This may be breathed in, depositing radium in the lungs and causing irradiation of the tissues lining the lungs. This problem was not recognized in earlier days, and there are instances, the most famous and well known of which is in Colorado, where building contractors used mill tailings as fill for house foundations. This resulted in high radon levels within the houses which are considered an unacceptable risk. Strictly speaking, of course, the mining and milling operations do not produce any *new* radioactive materials; they merely bring it up from underground. Nevertheless the waste can be disposed of safely by covering the "tailings" with about 20 feet of earth, which reduces the radon gas emitted to levels typical of ordinary soil. There are now federal regulations which stipulate standards for the disposal of mill tailings.

The most serious health hazard for the miners who dig out the uranium ore involves the inhalation of radon gas. Since uranium occurs in nature in association with radium and its decay series, quantities of this gas are continuously being produced. The accumulation of radon gas is not a hazard in the open strip mines; however, underground mines must be flushed with large volumes of air to reduce the radon to safe levels for the miners. Breathing radon and uranium dust is a serious occupational hazard, and uranium miners have, in the past, suffered from a high incidence of lung cancer as a result of working in poorly ventilated mines. This was particularly apparent in some areas of Colorado. Nowadays, underground mining is strictly controlled to limit this exposure. However, these risks must not be considered in a vacuum. All mining operations are notoriously hazardous, with an accident rate higher than for any other major industry. It is quite clear that the total cost in human suffering which results from the mining of nuclear fuels is considerably less than that involved in coal mining. This is a result of the small amount of uranium that is required to generate a given quantity of electrical energy, compared with the tonnage of coal used in a conventional plant to generate the same quantity of electricity. For example, the annual operation of a 1,000-megawatt coal-fired plant requires over 2 million tons of coal, versus some 175 tons of uranium which would require the mining of

80,000 tons of uranium ore. The annual toll from coal mining far exceeds that from any other kind of mining; every year on average there are several hundred fatal injuries and several thousand non-fatal injuries in the coal mines of the United States alone. It is a sobering statistic that by 1980, one hundred thousand men have died underground this century in U.S. coal mines—and the first woman!

The cost in human suffering from pneumoconiosis (lung disease) in coal miners must not be underestimated, either, since it involves hundreds of thousands of men. Figure 9.2 (p. 221) shows an x-ray of the lungs of a Welsh miner ruined by "black dust." Those who spent their childhood days in a coal mining area can translate these x-ray pictures into the familiar image of middle-aged men, debilitated by a hacking cough, thin and pale from lack of oxygen. These are not projected possible hazards; they happen every year to thousands of miners in every major coal field in the world. Mining coal has roughly 12 times the accident rate of mining and milling uranium, for the same amount of electricity generated. Chronic disability associated with coal mining is about 26 times as great as that of uranium mining for the same amount of electricity generated. The burning of coal also produces enormous amounts of air pollution which has been shown to be harmful to human health and damaging to the environment.

The final step in the nuclear fuel cycle is the conversion of the uranium concentrate to the gas uranium hexafluoride. This step is necessary because of the need to enrich the uranium fuel, that is, to increase the proportion of fissionable uranium-235, which in nature occurs as only one part in 140. As previously explained, some types of reactors, including the light water reac-

2 million tons of Coal

80,000 tons
of uranium ore

175 tons of
uranium fuel

FIGURE 7.17. Comparison of the relative amounts of coal and uranium required to operate a 1000 megawatt electricity generating plant for one year.

tors used in the United States, require 2 to 4 percent of uranium-235 in their fuel in order to operate. This enrichment process involves exploiting the difference in mass between fissionable uranium-235 and the heavier non-fissionable isotope uranium-238. This difference is best exploited with the isotopes in a gaseous state by passing them successively through porous filters. The only potential radiation hazard of a gaseous diffusion plant is from the leakage of the uranium hexafluoride gas, which can be kept to very low levels through careful technology, making escape from the plant extremely unlikely.

3. *Routine releases of radioactivity.* During the normal *routine* operation of a light water reactor, there are two principal pathways by which radioactivity can leak out and result in radiation exposure to the general public. The first is by the escape of radioactive gases; these may be fission products permeating through the fuel cladding which includes isotopes of Xenon, Krypton and Iodine, or produced by intense neutron bombardment of the air, which results in isotopes of nitrogen, oxygen and fluorine. Most of these pose no conceivable threat to the public because, either they have very short half-lives and decay away quickly or, as in the case of ^{131}Iodine, they are easily trapped and filtered out in the gas clean-up system or are dissolved in the water cooling system. The exceptions are the noble gases, Xenon and Krypton, which leak from perforations in the fuel cladding and therefore vary from reactor to reactor. These so-called *noble* gases are chemically inactive; the bad news is that they are consequently difficult to absorb on filters or by chemical reactions. However, the good news is that, being inactive, they tend not to be retained in the body; they are breathed in and breathed back out again. Two isotopes of Krypton are produced. Krypton-85, present in greatest abundance, has a long half-life of almost 11 years, but decays to a stable nonradioactive substance. Krypton-89 is of more concern, since it decays to a radioactive isotope of strontium, a solid, which is chemically similar to calcium. However, since Krypton-89 constitutes only a relatively small part of the gaseous release from a reactor, and since its half-life is only about three minutes, it is significant only close to plant boundaries. Six different radioactive isotopes of Xenon are produced as fission products in the reactor fuel. At high temperatures they can permeate the cladding of the fuel elements. However, they do not travel far from the reactor and are not an important factor.

The second pathway by which radioactivity from reactors can reach man is via the cooling water, which may contain fission products as a result of being in contact with the fuel elements. Before the water is released, an elaborate clean-up procedure is followed, and the water is stored for a period to allow for the decay of short-lived radioactive species. It is then diluted with fresh water before being discharged. Typical long lived fission products which can be present in tiny amounts include Strontium-89, Strontium-90 and Cesium-

137. Traces of radioactive cobalt may be present from neutron activation of the structural materials in the reactor. The level of radioactivity is extremely small; for example, it has been calculated that one could swim in the water continuously night and day for a year and receive only 0.1 microsievert (10 microrem) total body irradiation from the dissolved isotope—during which time of course one would have received 1 millisievert (100 millirem) from natural sources! There are some more subtle considerations, however, since it is known that the food chain of marine life can concentrate certain elements in a remarkable way. Thus, fish, shellfish, or seaweed eaten by man may contain a concentration of radioactive isotopes many times higher than the water in which they live and grow. This possibility has led to the routine monitoring of the products of the waters around nuclear reactors, but the radiation doses involved to the public turn out to be trivial.

Tritium is an isotope of hydrogen which contains two neutrons in its nucleus as well as the proton contained by the ordinary hydrogen atom. This isotope has a half-life of over 12 years and emits a weak β-particle. It is present in appreciable quantities in the liquid waste from reactors, which adds to that already present in the environment naturally from cosmic ray bombardment of the atmosphere, not to mention a contribution from nuclear weapons tested in the 1950s and 1960s. When taken into the body as tritiated water, it is diluted with all of the ordinary water in the body, and excreted again within a few days. It is not considered to be a health hazard at these extremely low levels. The controlled release of radioactivity into the air, and liquid effluents, was the subject of bitter controversy in the United States in the 1960s, but is rarely heard of today. The sequence of events as they happened is well worth reviewing. The agency regulating all nuclear reactors in the United States at that time was the Atomic Energy Commission (AEC), whose policy on radioactive emissions was "as low as practicable." The critics of nuclear power argued that this policy was simply not good enough, even though they conceded that in practice the dose to the public was vanishingly small. The objectors had a point, because after all what is "low" and what is "practicable"? The only radiation dose limits written into public documents is the level of 1.7 millisievert/year (170 millirem/year), the limit allowed to large numbers of the public. Technically, at least, the AEC could legally have irradiated every man, woman and child in the nation to a dose equal to this amount. However, scaremongering articles soon appeared in the popular press describing in gory details the health hazards (especially the risk of cancer) that might result as a consequence of this policy.

In response to this pressure, the United States Atomic Energy Commission introduced the rule that the dose at the boundary fence of a power reactor should not exceed 50 microsieverts/year (5 millirems/year). It is worth noting that a similar dose is accumulated during one transatlantic trip per year in a jet, or by moving from a wood to a concrete house for a few weeks of each year! Furthermore, not too many people live on the boundary fence of a reactor, since such devices are usually built in sparsely populated areas. Even when all projected power reactors are built in the United States, the average dose to the population as a whole will barely reach one-tenth of the figure at the boundary fence. The change in the rules has satisfied even the most vocal of critics (at least on this point!) and documents what has always been practiced, namely, that the radioactive releases should be as low as practicable. Before leaving this topic, it should be pointed out that this miniscule release of radioactivity replaces the clouds of black smoke which billow from oil- or coal-fired stations. The hazards to public health from these noxious fumes, containing soot and dust particles as well as toxic gases and corrosive chemicals, far exceed the worst estimate that can be made of the effects of routine emissions from nuclear reactors.

4. *Processing and disposal of nuclear wastes.* Operation of a nuclear reactor produces vast quantities of a wide variety of radioactive isotopes, some of which have very short half-lives of only a few seconds while others have long half-lives of many hundreds of years. This radioactivity consists of the fission products resulting from the break-up of uranium and also the production of materials such as plutonium, americium and curium, which are generated in the reactor by the capture and retention of neutrons. While safely contained in the reactor, these radioactive materials pose no problem, but they must be disposed of when the fuel is exhausted. There are two alternative plans. First, the fuel elements may be opened and reprocessed to recover the uranium that is unused and to separate the plutonium that is generated during the operation of the reactor. Second, the fuel elements may be removed, placed in shielded containers and stored unprocessed in well-guarded and shielded repositories. *place where things are stored.*

Reprocessing. The reprocessing of fuel elements requires a special reprocessing plant; installations of this kind are operated by the Department of Energy in the United States at Hanford and at Savannah, and there are similar installations in Europe, for example, Windscale in England and Le Havre in France. There are three basic products of a reprocessing plant. First, the recovered uranium which can be recycled and used again in a reactor fuel element. Second, plutonium which could also be used for reactor fuel but has a greater value for the production of nuclear weapons. Third, high level wastes, the fission products of uranium, must be safely disposed of.

Up to the present time, the motive for essentially all of the reprocessing that has been carried out has been to obtain plutonium to produce nuclear weapons; however, in the future any enlightened long-term nuclear power program would require plutonium to be used as a reactor fuel.

    The first step of the reprocessing process is to break up the fuel elements and dissolve them in strong acids, which permits the gases and volatile reactive fission products to be released. These gases are filtered and scrubbed with water which removes essentially all of the radioactive iodine但but allows a great deal of radioactive tritium and essentially all of the Krypton-85 to escape into the atmosphere. As previously described, Krypton-85 has a half-life of about 10 years and is chemically inert, so that it is breathed in and breathed out again. It is therefore thought that low levels of this isotope are essentially nontoxic, but it must be a cause of some concern if the total amount of Krypton-85 in the world environment were to increase substantially. It is interesting to note that the one-time release of Krypton that occurred in the Three Mile Island accident, which received so much publicity, is equalled about every 10 days in the routine operation of the European reprocessing plants in France and in England!

Tritium released in reprocessing plants is another area of some concern. It is released partly as a gas, while some is retained in the liquid waste. Since tritium is an isotope of hydrogen, it readily exchanges with the body's normal hydrogen but is readily eliminated. Again, it is considered that low levels of tritium do not pose a significant risk, but it would be of some concern if the amounts in the atmosphere were to increase by a large factor.

The uranium and plutonium are separated from the remaining radioactive wastes in a process that is relatively simple in principle, but the separation is made difficult by the extremely high radiation levels involved. Enormous quantities of highly radioactive waste products collected in the reprocessing plants are stored in vast tanks. In the United States program, these high level wastes are stored at Savannah River in Georgia. On a laboratory scale, this liquid waste has been converted into a solid, greatly reduced in bulk, but this process has not yet been developed on a larger scale to cope with all of the waste products of the nuclear power industry. It would be highly undesirable for there to be any leakage of this waste material, so as a matter of policy, waste management is directed towards solidification of all high level wastes for final long-term disposal. Solid wastes could then be buried deep in stable geological formations and should be safe over long periods of time.

Most of these fission products have a half-life which does not exceed 30 years; after a period of 700 years, less than one ten-millionth of the original activity remains and by then the activity of the stored waste is comparable to natural deposits of uranium or thorium ores. For this reason 700 years is often taken as the limit of practical concern for this category of radioactive waste. One of the major problems of waste disposal at the present time is choosing and identifying suitable sites for final burial; the problem is more political than technical since there is a very natural tendency for anyone living near a proposed burial site to voice their objections loudly! We would all like waste products of all sorts to be buried in someone else's backyard! While the amount of radioactive waste is large in terms of activity, the actual volume and the amount of land required to store it is relatively small. A point frequently raised is that long-term storage represents a strong commitment, not only of our present society and its descendants, but perhaps even a different society at some date far into the future. This is, of course, true. Although it poses no great hazard to the public at present, radioactive waste may cause a problem for some future tenants on earth, even though it is likely to be a trivial risk compared with other industrial pollutants, including toxic chemicals and the build-up of carbon dioxide in the atmosphere by the combustion of fossil fuels.

For many years Britain has followed a policy of dumping high level radioactive waste far out to sea. Several thousand tons of this waste are disposed of in this way each year. After international consultation, a site is chosen where the water is deep, where there are no strong up-currents and where commercial fishing is not practiced; for example, a dumping site used at present is several hundred miles off Lands End, the southwestern tip of England. By design, the containers break up over a period of time, and the radioactive material is diluted with vast quantities of sea water, which already contains a great deal of natural radioactive material. Computer models indicate no appreciable hazard from this practice, but there are loud and frequent shouts of protest from conservationists and environmentalists. Dumping in the sea has been considered and rejected in the United States, because once radioactive material is dumped, control over it is lost. Current American philosophy is to store radioactive waste in "tombs" or repositories, where it is accessible and from which it can be retrieved once a permanent method of storage is agreed upon.

In Great Britain, low level radioactive wastes are pumped out to sea with stringent control of the amounts dumped and careful surveillance of marine life to detect possible effects on human health. In fact, the story of what happens is quite fascinating. The British fuel reprocessing plant is at Windscale in Cumberland on the northwest coast of England. Radioactive waste is piped far out into the Irish Sea, but a tiny fraction of it is washed back onto the shore by the currents and tides. It turns out that the human population subject to the greatest risk consists of about 26,000 people in South Wales, hundreds of miles to the south. These are individuals who eat "laverbread" — a local delicacy made from an edible seaweed *Porphyra umbilicalis*, which is the gastronomic specialty of the Gower coast. When cooked, it is a dark green, soggy mess, not unlike the looks of spinach. Unkind critics, surprised by its appearance, have described it as the only edible cow-pat in the world! They have obviously never tried it dusted with oatmeal and fried with bacon and eggs for breakfast; it is in fact quite delicious! The lovers of laverbread in South Wales are, for the most part, under the impression that the seaweed is collected fresh each day on their own shoreline around Swansea, Mumbles and the beautiful Gower Peninsula. In fact, demand cannot be met from

FIGURE 7.18. The approximate area in South Wales where laverbread is regularly eaten. This delicacy is prepared from seaweed harvested on the northwest coast of England where radioactive waste is washed ashore from the Windscale fuel reprocessing plant.

local sources, and most of the seaweed is brought by train from northwest England where it is harvested in the vicinity of the reactor waste discharge areas. Consequently, it contains traces of radioactive ruthenium and cerium. The most significant radiation dose to laverbread eaters is to the lower large intestine, where as much as 11 millisievert/year (1.1 rem/year) can be accumulated by individuals who consume the delicacy regularly. This does not break the regulations of The International Commission on Radiological Protection concerning exposure of the general public and probably amounts to a very small risk. At all events it seems to be accepted without complaint by the inhabitants of South Wales; a similar state of affairs would not be tolerated in the United States, where the public is much more sensitive to such issues and much more likely to launch an organized protest to the authorities.

So far, all that has been said about reprocessing plants concerns routine operations. In addition, there is always the chance of a serious accident, which would, however, be on a relatively small scale, with the consequences confined to the building in which it occurred. It is difficult to conceive of a serious consequence to the public. In the 40 or so years since the first reactors were built, a few individuals have lost their lives in the performance of their duties, one of whom was in a reprocessing plant. However, as pointed out in an earlier chapter, our relative ignorance of the effects of large doses of radiation on humans stems from the incredible safety record of the whole nuclear industry, as a result of which so few lives have been lost. An ordinary factory, or a conventional coal mine, is a much more dangerous place to work than either a nuclear power reactor or a reprocessing plant.

STORAGE WITHOUT PROCESSING

Reprocessing of commercial nuclear fuel elements has been discontinued in the United States because of the concern that plutonium, a by-product of reprocessing, could be diverted to produce nuclear weapons. It is argued that plutonium could be seized by terrorists or secretly transferred to other governments in amounts large enough to make a bomb. By simply storing the unopened exhausted fuel elements directly in deep burial sites, the possibility of the proliferation of nuclear weapons is minimized and all of the hazards of reprocessing avoided. The supply of fuel to conventional reactors is reduced. The fuel elements from commercial nuclear power reactors are stored under water to large storage pools at the reactor sites or at commercial spent fuel storage sites in Illinois. These are clearly temporary sites pending the choice of deep burial sites where the fuel elements may be stored long term.

In summary, nuclear waste from power reactors can be disposed of safely in two ways, either by storing the spent fuel elements without ever opening them, or by reprocessing to extract the plutonium and uranium and concentrating the fission product wastes in solid form for deep burial. This issue has not yet been settled in the United States, and is political rather than scientific. It involves the balancing of the risk of weapons proliferation because of the availability of weapons-grade plutonium that can be separated, against the gain of the availability of a vastly increased supply of usable fuel for nuclear reactors in the future.

5. *Transportation of radioactive waste.* At some stage, a decision will have to be made as to whether nuclear power plants should be distributed uniformly around the country or clustered together in "nuclear parks." This decision will influence the extent to which radioactive wastes and used fuel elements are transported between power reactors, reprocessing plants, and storage vaults.

In evaluating transportation hazards, it has been estimated that the driver of a truck carrying spent fuel elements is the most likely person to be injured.

The chances, however, are between 2,000 and a 120,000 times greater that the injury will be from the effects of the crash itself rather than from some radiation cause! Sizable amounts of radioactivity have been transported over American highways for a number of decades, without known harm to the public. Spills have occurred occasionally, of course, but these have involved small amounts of radioactivity that are a nuisance to clean up but no threat to public health. In the past 25 years about 250 accidents or incidences in transportation have been recorded in which radioactive materials were involved. None caused death or injury of people as a result of the radioactive nature of the material. However, the skeptics believe that this record will not continue, and who can argue with them?

Radioactive wastes are stored in containers that are designed to withstand the impact involved in a severe accident. There is no chance of an explosion of the cargo sufficiently severe to spread the material over a large area, and the confinement of any radioactivity that might be freed to the immediate environment will render its potential for public health hazards small indeed. As far as danger to the public is concerned, it is negligble compared with the transportation of gasoline, naphthalene and other explosive and inflammable products.

6. *Reactor accidents.* When all is said and done, one of the most serious causes of concern as the number of reactors in use increases is the remote possibility of an accident. There is an old proverb that says: "If an event *can* happen, then given time it *will.*" With so much radioactive material packed into so many reactors, some people fear that sooner or later one of them will malfunction, and there will be a disaster. What are the chances of this happening? As of the 80s, there were a total of over 2,000 reactor-years experience throughout the world involving commercial plants producing electricity, i.e., the equivalent of one reactor operating continuously for 2,000 years. During this time there have been "minor incidents," as in any industry, but no accident in which the release of radioactivity injured personnel working in the plant, let alone the general public. This is an enviable safety record, and indicates that the probability of a severe accident is very low. On the other hand, it cannot be concluded from this limited experience that accidents are not possible and that they will never occur in the future.

At this point it is instructive to review the accidents that have occurred over the years:

The Windscale Accident—1957. This accident occurred in England and involved an air-cooled reactor with a graphite core. The intense neutron flux within this type of reactor leads to a buildup of energy in the graphite, which must be routinely released by allowing the core to warm up; this is achieved simply by temporarily restricting the cooling supply. On this occasion, part of

FIGURE 7.19 (a) and (b). Radioactive spent fuel elements are specially packaged before they are transported over public highways to storage dumps. These pictures illustrate a "crash" staged to test the efficacy of the packaging. (Top) A truck carrying the radioactive material is involved in a high-speed collision with a train on a level crossing. (Bottom) The truck is destroyed, but the specially designed container for the radioactive spent fuel elements is recovered intact and undamaged. (Courtesy of Sandia National Laboratories, New Mexico, USA.)

the core overheated resulting in a fire. The fire was put out by flooding the reactor with water but not before large amounts of radioactive material had been released. The situation was aggravated by the fact that this early type of reactor was not housed in a containment building designed to limit the effects of an accident of this sort—a design feature which was to assume great importance in the later accident at Three Mile Island. Large amounts of radioactive Iodine-131, Cesium-137 and Strontium-89 were released, together with smaller amounts of other isotopes. The prevailing westerly wind blew the radioactivity over the English countryside and across the North Sea to continental Europe! Not only was there no containment building enclosing this reactor, but there also were no charcoal filters to trap the radioactive Iodine-131, both standard features in commercial light water reactors in the United States. Surveys indicated that the milk of cows grazing over a considerable area was contaminated with Iodine-131 and for over a month was deemed an unnecessary risk for human consumption. (This was the most serious reactor accident ever in terms of the quantity of radioactivity released—at least in the Western world.) Extensive environmental measurements were made during and after the release, which permit an estimate to be made of the radiation doses to the exposed individuals in the local population. Under the plume from the reactor stack in the downwind direction, the dose-rate from gamma rays reached a high of about 40 microsieverts/hr (4 millirem/hr), and over the next few weeks an individual in the area might have accumulated up to 200 microsieverts (20 millirem) from gamma rays given off by the fallout.

In 1983 the collective effective dose equivalent commitment from the release was published as 1,200 person sieverts (120,000 man rems). The route of exposure which contributed the most was the consumption of contaminated milk. Iodine-131 was the most important radionuclide, contributing nearly all of the collective dose to the thyroid and the largest part of the collective effective dose. In the longer term, the contribution of Cesium-137 to the collective effective dose via external irradiation from ground deposits and the ingestion of contaminated foodstuffs became more significant. This leads to the estimate that possibly 11 cases of cancer could have been induced in the millions of individuals exposed. This estimate, of course, assumes a proportionality between cancer induction and radiation dose, down to doses which are a fraction of natural background levels.

Idaho Falls—1961 (the SL-1 incident). This accident involved a small experimental boiling water reactor, designed to produce electrical power and heat for military installations located in places remote from the usual power sources. Being an early design and located in a relatively deserted area, it did not have a containment building. A crew of three had been assigned to start up the reactor after a long shutdown, and for some unknown reason, a local excursion occurred, producing a steam over-pressurization in the reactor vessel. The consequences of this steam explosion resulted in fatal traumatic injuries to two persons and were contributory to the third person's death, although he also had received a large exposure due to the radiation from the release of fission products from a damaged fuel element. Rescue workers measured a dose-rate in excess of 10 Sievert/hr (1000 rem/hr) within the building; a 30-minute exposure at this rate would be lethal due to failure of the blood forming organs, with death occurring 30 to 60 days later, while an exposure of an hour or two would be lethal due to failure of the lining of the intestines, with death occurring 9 to 10 days later.

This was a tragic accident, involving loss of life; it could have been worse had not most of the radioactivity stayed within the building, despite the fact that it was not a containment building as used nowadays. But half a mile from the boundary fence, the dose-rate was only 20 microsieverts/hr (2 millirem/hr). There were, in fact, no significant consequences beyond the reactor building itself.

The Tennessee Valley Authority's reactor at Brown's Ferry—1975. This installation consisted of two 1,000 megawatt boiling water reactors. Following maintenance in one of the reactors, a workman using a candle flare to check for leaks between the control room and the reactor containment area ignited the insulation of the electrical wires. There was a significant spread of the flame which was sucked into a crack, igniting the plastic sealant which smouldered and burned and spread unnoticed for some time because it was behind the wall of the control room. The reactors were shut down immediately, but as explained earlier, water is still needed to cool the core and remove heat produced by the decay of residual radioactive materials; this became increasingly difficult as one by one, control cables were being burned and their function lost. Because of the backup systems and redundant controls, the coolant level was maintained and eventually the fire was put out by flooding the area with water. In spite of the destruction of much of the control circuitry, the reactor core was undamaged, no one was injured and no radioactivity was released. The only consequence of this serious accident was a financial loss to the Tennessee Valley Authority. This is a good example of the strategy of multiple redundant control systems, which are a vital part of the design of modern reactors.

Three Mile Island—1979. Three Mile Island became a household word as a result of an accident that occurred in the early morning hours of March 28th, 1979. It is the site of two Pressurized Water Reactors (PWR) operated by Metropolitan Edison Co. and is located on an island in the Susquehanna River close to Harrisburg, Pennsylvania.

The accident occurred because of a bizarre combination of human error and mechanical malfunction. It is hard to read through the various reports and distill out the essence of what happened, but the critical sequence of events appears to be as follows:

1. Maintenance men inadvertently caused the shutdown of a pump which circulated condensed steam from the turbine to the steam generator. *Human error number one.*
2. An auxiliary supply should have cut in automatically, but valves designed to be open were closed. *Human error number two.* Because of a rise in pressure in the reactor, the control rods were inserted automatically to shut down the reactor and a relief valve opened to release excess pressure. These functions operated automatically and by design.
3. The relief valve failed to close when normal pressure was resumed; *mechanical failure number 1.* This should have been no problem because there are ways to close the valve manually *except:*
4. The indicator on the control panel showed the valve to have closed when in fact it was open; *mechanical failure number 2.*
5. For over 2 hours steam and water were being lost; meanwhile the operators, misled by the faulty indicator, did not realize what was happening and did not initiate other cooling procedures. *Human error number 3* (it is perhaps a harsh judgment to label this an error).

As a consequence of this unlikely combination of multiple human errors with multiple interrelated mechanical malfunctions, the level of cooling water fell and the reactor core was partially uncovered, which resulted in partial melting of the cladding of the fuel elements. As a result, there was a major release of radioactivity into the steam and cooling water, some of which escaped from the reactor building, releasing radioactivity to the outside when the water was being transferred to an auxiliary building. It must be admitted that the containment building was breached, albeit to a minor extent.

During the accident, about 10 percent of the radioactive noble gases in the reactor, mostly Xenon and Krypton, escaped into the atmosphere. To put this into perspective, the noble gases released over a period of a few weeks were equal to the amount that would be released under normal operating conditions in about 10 years. Little radioactive iodine was released because most of it was removed in the charcoal filters of the gas clean-up system or dissolved in the water and remained in the plant. Essentially all nonvolatile fission products, the vast bulk of the radioactivity, remained within the containment building.

The commission appointed by the President, as well as that appointed by the Nuclear Regulatory Commission, agree that the amount of radiation resulting from the radioactivity released was too small to produce any measurable harm to the nearby population. This calculation is relatively simple to perform. Two million people lived within 50 miles of the plant. Various agencies, task groups and committees made estimates of the collective dose equivalent received by this population, which varied over a wide range from 5 to 100 person sieverts (500–10,000 man rem). This is the product of the number of people exposed and the average dose to which they were exposed; the technical term for this quantity is the collective dose equivalent. Of the many dose estimates made, perhaps the most credible are those of the Ad Hoc Interagency Committee at 33 person sieverts (3,300 man rem), and the President's Commission Task Group at 28 person sieverts (2,800 man rem). The *average* dose equivalent therefore, to the 2 million people living within 50 miles of the plant works out to be only 15 microsievert (1.5 millirem). In fact, of course the doses received were extremely nonuniform with about a dozen individuals receiving close to 1 millisievert (100 millirem), while the bulk of the population received less than 10 microsievert (1 millirem). Taking a value of 30 person sieverts (3,000 man rem) for the collective dose equivalent, midway between the most credible estimates, and using the cancer risk estimate of 1 death per 100 people exposed to 1 sievert (discussed in chapter 3), we can calculate the number of cancer deaths expected from the TMI to be:

$$30 \times \frac{1}{100} = 0.3$$

Whichever dose estimate is used, the number of cancer deaths expected does not exceed one. It should be noted that of a population of 2 million people in the United States, about 16 percent, or 320,000 will die of cancer from natural causes. It is clear, therefore, that the effect of the TMI accident on the local cancer incidence will never be seen against the natural incidence.

Postscript on reactor accidents in the wake of the Three Mile Island. A great deal has been written about the Three Mile Island accident, and now that a period of time has passed it is possible to reflect upon it and summarize some useful conclusions.

First, the accident was the worst ever in the United States involving a nuclear reactor used for the generation of power. If this is the worst, and it was relatively trivial by any standards, then it bodes well for the general safety of reactors. However, it does indicate that accidents can happen. If a trivial accident happens with low probability then presumably a serious accident *could* happen, even if the probability is remote.

Second, in spite of a mechanical failure which started the sequence of events at Three Mile Island, followed by an incredible series of bungling errors by the operators, the eventual consequences were not very serious. The performance of the operators could hardly have been worse if they have been trying deliberately to sabotage the reactor. In spite of all these foolish actions, the consequences of the accident were contained and no one was hurt, much less killed.

Third, one worrisome feature is that the statistics often quoted for the thousands of reactor hours accumulated without a major accident are comprised largely of experience during the time when reactors were in the experimental phase and when they were operated directly or indirectly under the supervision of the Atomic Energy Commission in the United States, and comparable bodies in other countries. These were very much elitist institutions, where well-trained individuals operated in a relaxed atmosphere with generous funding and where no attempts were made to jeopardize safety in the interests of economy. Now that power reactors have passed from the experimental to the routine, and now that their operation has passed from the elite Atomic Energy Commission to power companies, who must make a profit and who have difficulty finding people with adequate backgrounds to train for their operator positions, one wonders whether the same high standards of safety can be maintained. The operation of a conventional commercial power producing reactor is tedious and boring and largely in the hands of people who are not very highly trained; this must be a cause of at least some concern.

Fourth, one good feature of this accident is that the publicity and the lessons learned at TMI have caused the United States Nuclear Regulatory Commission (USNRC) to increase their surveillance of these reactors and the qualifications of their personnel.

CATASTROPHIC ACCIDENT

The incredible past safety record of nuclear reactors in general, and commercial power reactors in particular, does not offer any guarantee that a serious accident will *never* occur in the future. Based on experience accumulated to date, attempts have been made to estimate the likelihood of an accident and to predict its outcome. The most exhaustive study was the Rasmussen report, published in the 1970s. The study considered the most likely causes of accidents and also studied unlikely combinations of rare events to create a worst possible accident. The probability of such events was calculated as well as the consequences in terms of the release of radioactivity, property damage and loss of life.

It is clear from the Rasmussen report that the principal potential hazard for an accident at a power plant is the release of radioactive materials, not an explosion in any way comparable to a nuclear bomb. It further emphasizes that melting fuel elements is the only way in which significant amounts of radioactivity can be released. The Three Mile Island accident demonstrated that even fuel cladding damage due to reaction with steam at high temperatures would not result in the release of substantial quantities of radioactive materials. In either case, damage to the fuel elements results from the failure of the cooling system due to human error or mechanical malfunction. It is a commonly held belief, partly as a result of the 1979 movie *The China Syndrome*, starring Jane Fonda and Jack Lemmon, that cooling failure would cause the core to melt into a fiery ball that would melt through the pressure vessel, through the containment building, and release vast amounts of radioactivity. This concept led to an exciting and dramatic movie, but it bears little resemblance to fact, as the events at Three Mile Island illustrated. At TMI the cooling system failed, but there was no major disaster, despite the release of some gaseous radioactive isotopes. The Rasmussen report estimates that the worst possible accident, involving a major breach of the containment system leading to the loss of several thousand lives, would be expected to occur once every 10 million years in the United States, which has about 100 commercial power reactors in operation. If society were unwilling to accept risks of this order, we could never build a dam to store water in a reservoir, after the accident in Pennsylvania in 1889 that claimed 2,000 lives, nor could ocean-going liners be considered after the sinking of the Titanic in 1912 which cost 1,500 lives. In summary, then, the consequences of a worst possible nuclear accident are severe, but no worse than other major natural or man-made disasters, while the likelihood of its happening is vanishingly small.

Table 7.1. Number of fatalities and injuries expected per year among the 15 million people living within 20 miles of 100 nuclear reactors

Accident type	Fatalities	Injuries
Motorcar	4200	375,000
Falls	1500	75,000
Fire	560	22,000
Electrocution	90	—
Lightning	8	—
Reactor	0.3	6

Source: Taken from the Rasmussen report on reactor safety

Table 7.2. Probability of major disasters—man-made and natural

Type of event	Probability of 100 or more fatalities
Natural:	
Tornado	1 in 5 years
Hurricane	1 in 5 years
Earthquake	1 in 20 years
Man-caused	
Airplane crash	1 in 2 years
Fire	1 in 7 years
Explosion	1 in 16 years
Toxic gas	1 in 100 years
Reactor (100 plants)	1 in 10,000 years

Source: Taken from the Rasmussen report on reactor safety

THE PERCEPTION OF RISK

Very few of us are logical in the way we face up to risks in our lives. For example, most people are much more afraid of flying in an airplane than of driving in a car. It makes no sense, since we all know that the safety record of the airlines is excellent and that driving a car is much more dangerous. This knowledge does nothing to make us rational concerning the relative hazards of driving and flying. The reason is easy to understand. When a plane crashes, it involves a frightening disaster—smoke and wreckage and hundreds of charred bodies littered over the countryside. It doesn't happen often, but when it does, it is staggering and spectacular. By contrast, the death toll on the roads goes on continually. Every hour of every day people are killed or mutilated in motorcar crashes, one or two at a time, and it seldom makes the headlines. In the end far more people are killed on the highways than in airplanes—and however you express the statistics it is safer to fly than to drive. But this does not deter our prejudice, based on the spectacle of a big crash.

In a way it may be the same for the rival methods of power production. Nuclear power reactors *do* have the potential of a catastrophic accident—this cannot be denied. The risk is vanishingly small, but if there are hundreds of reactors around the world, eventually there is likely to be a disaster, killing many people. It will not happen often, perhaps once every 10,000 years. Power stations fired by coal and oil can never cause a disaster of such proportions. There is no way it can happen. But every day they pump out soot and smoke and all the associated toxic gases. The health of the public in their vicinity is adversely affected, and every year some people die as a consequence. There is no question that producing electricity from the atom will kill fewer people than the use of coal and oil, but the pattern will be different, just as air fatalities differ from deaths on the roads. The choice is an occasional disaster, possibly killing many people at once, or the gradual day-by-day killing of one at a time. We are used to the latter—the former still shocks us and worries us. Despite all of the reasoned arguments, this is the basis of a remaining doubt in our minds about nuclear power.

7. *The homemade atom bomb.* As the civilian nuclear industry grows throughout the world, and there is a proliferation of reactors routinely used to generate electricity, it is argued that there is an increased potential for fissile material to be stolen and diverted to clandestine purposes. The risk of nuclear blackmail or sabotage is thereby increased. An eccentric individual or an organized gang of terrorists could possibly steal enough fissile material to make an atom bomb, and they could threaten to use it if their outrageous demands were not promptly met.

Critics count this as yet another cause of concern about the desirability of expanding nuclear technology. The present water-cooled reactors use a fuel which is enriched to only a few percent of uranium-235, and this cannot support a nuclear explosion. Furthermore, to increase the uranium-235 fraction to a point where it would explode is no job for the backyard scientist; at present it is not feasible without a big national effort. However, within the next few years substantial amounts of plutonium isotopes will be created in commercial reactors from uranium-238. Recycled plutonium is not the first choice for a high-quality atom bomb, but while the explosive performance would be quite unpredictable, it would not be out of the question.

The appropriate agencies in both the United States and Europe, as well as the International Atomic Energy Agency (IAEA), all recognize this problem and have taken steps over the years to stop fissile materials falling into the wrong hands. The security surrounding nuclear power reactors is beginning to approach that characteristic of the military who have similar concerns about diversion. It does, at least, indicate that the authorities recognize the problem involved with the civilian use of these materials. However, it is evident that theft or diversion of biochemical warfare agents inherently has the same potential for mischief, so that the problem is not new.

Just how big a hazard is it to the man in the street that an eccentric crank or well-organized gang could steal enough material to make a bomb? The first, and in the end quite effective, safeguard is that enriched fuel is extremely expensive and merits careful guarding like precious metals or scarce minerals. It would not be an easy matter to steal in the first place. Nor could the average person handle it safely and construct a workable bomb—though there must be several thousand scientists and engineers in most countries who could. But it must be conceded that an equal number of individuals could just as well divert the expertise of germ or chemical warfare to produce far more devastation than any bomb. If it comes to that, a few dozen cans of vichyssoise soup, infected with botulinus and strategically deposited in the water supply of a large city could kill far more people than any atom bomb. It may not have quite the same emotional appeal, but it would work with exquisite efficiency.

When all is said and done, there is no shortage of possible weapons for evil men to kill, terrorize or blackmail society. A homemade atom bomb is just one more added to the long list already available. It is rather less practical than most, and not all that effective. Our Western society, so relatively open and free, is particularly vulnerable and susceptible to acts of terrorism. It is remarkable that such disastrous events are not more common than they are.

8. *Nuclear proliferation.* In the long run, perhaps the most serious and legitimate concern arising from the increased use of nuclear power is the

spread of the associated nuclear technology. This carries with it the clear responsibility and indeed the likelihood that it will lead to an increase in the number of members of the "nuclear club," i.e., nations with access to nuclear weapons. The way in which the process operates was demonstrated clearly by India in the early 1970s. The Canadian government generously provided a CANDU heavy water reactor as part of their aid program, to help produce the energy needed to develop India. The Indian government elected to operate the reactor in a mode to produce maximum weapon grade fissionable material, i.e., plutonium, so that they could produce a nuclear weapon. India demonstrated to the world that, despite high sounding holier-than-thou criticism of others, they were quite ready to cheat as soon as they had access to nuclear reactors provided by the generosity of Western countries with the clear understanding that they were for the generation of power. The desire to have access to nuclear weapons is clearly strong in the minds of many countries—either because of fear of their neighbors, a desire to conquer their neighbors, to bolster a fragile ego, or some combination of all the above.

The nuclear "club" to date has shown remarkable but sensible restraint for almost 40 years, having seen the destructive power of nuclear weapons demonstrated at Hiroshima and Nagasaki. One wonders whether similar restraint would be shown by leaders of military dictatorships, such as many in South America or Africa. Specifically, for example, one wonders if Argentina would have been tempted to use nuclear weapons, had they been available, when unexpectedly the Falklands War turned against them, and they were soundly thrashed by the British task force. Every effort must be made to keep nuclear weapons out of the hands of dictators, whose support depends upon a people of volatile temperament, and who lose their position of power and perhaps their lives when they stop winning! In this context the temptation to gamble all on a limited nuclear war might be irresistable. The unfortunate corollary of this argument is that if nuclear weapons are not to be allowed to spread, then nuclear reactors for power generation cannot be distributed to these nations either. The Western industrialized nations can utilize nuclear power to satisfy their substantial appetite for power, both to maintain their high standards of living and to maintain progress in science and technology, while further loosening the grip of OPEC. The increased use of nuclear power, therefore, must mean the increased use in the industrialized countries, which already have access to nuclear weapons. It would be folly to tempt the proliferation of nuclear weapons by encouraging the increased use of nuclear power in countries with unstable and unreliable political systems and with yearnings to join the nuclear club. Unfortunately the sale of nuclear reactors for power production is not controlled by any responsible body. As mentioned earlier, Canada has an active and well-developed nuclear industry and must export to remain viable. As a rule of thumb, one power station per year must be sold for an industry to remain solvent and Canada does not require that many at home. Consequently, a whole list of countries in the Middle East and South America, including Argentina, are negotiating to buy Canadian reactors. These Canadian-built heavy water reactors are bigger and more expensive over the short term than the smaller light water reactors built in the United States, but they produce plutonium very effectively and are the logical choice of a nation for whom power production is a consideration secondary to building a nuclear weapon.

SOLVING THE ENERGY PROBLEM

An energy crisis dawned on the Western world during the 1970s. For years, prophets of doom had forecast that it was inevitable, but no one paid much attention. Now that it has finally arrived, it is likely to be with us for a long time. There may be periods of respite during economic recessions, but the energy shortage must return to haunt us again and again. The course of action taken at a national level to meet the energy crisis will necessarily be conservative and essentially a compromise. This must be so because of the democratic process. All kinds of vested interests must be protected, jobs safeguarded, and at least lip service paid to considerations of conservation.

The solution of this problem exercises the minds of some of the world's most able scientists, and is raw material for countless "think tanks." I would be a monster of egotism to even suggest that I know how the energy crisis will be solved. What follows is a consensus of current thinking about the most likely ways in which the problem will be alleviated, particularly in the United States, and the role to be played by nuclear reactors.

Short-term answers to the problem must be found and implemented immediately; a great deal of careful thought must be given to long-range solutions.

SHORT-TERM SOLUTIONS

The keynote of the short-term solution must be conservation. Better insulated houses, smaller motorcars, less emphasis on running to and fro. It is already leading to a revival of the city centers, so that people can live closer to their work. A most important contribution would be a reordering of social priorities and a slowing down in the upward spiral of living standards and the accumulating of material wealth.

What oil there is can now be supplemented by less rich deposits, such as shale oil, that in the past were too expensive to harvest. The increased price of fuel, which is unlikely ever to be reversed, leads to an increased supply. Wherever feasible, oil and natural gas should be replaced by coal and nuclear power. Oil must be conserved for transportation, for automobiles and airplanes, for which no reasonable alternative fuels exist. There is no sense in using up limited oil resources to produce electricity in central generating stations. Coal and nuclear fuels are the logical replacements for this purpose. Existing oil-burning stations can be converted back to coal with a minimum of effort. New power plants must be based on coal or nuclear fuel.

Coal is dirty and produces a great deal of pollution. There is an obvious need for research and development to improve the efficiency of coal burning. Pollution levels must be reduced, particularly sulphur dioxide and soot particles, which when combined are killers in terms of lung disease. Relatively little research effort has been devoted to coal in the past. There seemed little point during the years of plenty, when oil oozed from the ground in great profusion with a minimum of effort. But now things are different. Coal is a national resource that the United States has in prodigious amounts. However, sufficient coal to meet the voracious appetites of electricity-generating stations can only be produced by opencast mining. Although known to have been carried out in Roman times, opencast mining began in earnest in 1941 to meet wartime needs. Since then, the industry has developed to a point where vast quantities of coal can be produced economically and profitably. The effect on the environment is described as "transitory," and so it is, but it is

FIGURE 7.20a. Aerial view of open-cast mining in South Wales, showing the giant slashes made in the landscape, and the devastating effect on the environment while operations are in progress. (Photograph by courtesy of Mr. T.J. Clement and The National Coal Board Opencast Executive.)

FIGURE 7.20b. The results of extensive reclamation as a result of which the natural beauty of the countryside is eventually restored. (Photograph by courtesy of Mr. T.J. Clement and The National Coal Board Opencast Executive.)

devastating while it lasts! Over the whole area to be mined, topsoil up to one foot thick is stripped by scraper and piled around the site in dumps, which are grassed over to improve their appearance. These mounds of earth also perform the added function of acting as a sound baffle to deaden the ear-shattering din of the equipment working night and day. Huge mechanical shovels then remove the rock and soil to expose the coal seams. Frequently, 20 to 30 tons of unwanted earth must be removed for each ton of usable coal. If the rock is hard, explosives are used to blast it out.

Opencast workings of this kind have the potential for all kinds of pollution and damage to the environment. Clouds of dust from the shale roads used to haul out the coal, noise and oil pollution from the mechanical gadgets, and floods caused when heavy rains fall on ground denuded of vegetation, all combine to make this form of mining very unpopular with residents of nearby areas. Worst of all is the fact that the giant slashes made in the landscape are an eyesore for a period of many years. It must be admitted that a great effort is made at restoration afterwards. The land is filled and regraded, the topsoil replaced, and trees or grass replanted. In many cases the landscape actually is "improved," or at least made more suitable for public parks or sporting facilities. But this takes time as well as money. In practical terms, an ugly scarred landscape for a 5 to 10 year period is the price society pays for the black diamonds extracted, while clouds of soot and smoke are the price paid for their conversion to usable energy.

Meanwhile more and more power will be generated by nuclear reactors, but the increase is likely to be very nonuniform from one country to another. It is interesting to note that countries with a vociferous antinuclear lobby, which are able to effectively close down their industry, happen to be those which have extensive coal reserves so that opencast or strip mining on a massive scale is a feasible short-term solution to the energy crisis. Most notable in this connection are the United States and West Germany. On the other hand, countries with less effective protesters against nuclear power—such as France and particularly Japan—have little coal and less oil in the first place, so that a massive turn to coal is not an option. In fact, one wonders what they would have done in the short-term if their antinukes had been as effective as in the United States. One can only speculate whether this apparent relationship between the availability of coal reserves and the effectiveness of the antinuclear lobby is coincidental. Britain does not fit into this pattern; they have substantial coal reserves but are steadily going nuclear anyway. This may be related to the fact that the nationalized coal industry in Britain is notoriously inefficient, making coal relatively expensive.

In the past, some have advocated a crash program in nuclear power, at least in those countries where it is accepted by the public, since it is so much cleaner and more acceptable ecologically than coal. This might have been possible in the sixties or early seventies, but is unlikely to happen now for several reasons. First, nuclear power stations cost more to build; they are competitive with coal and oil on a day-to-day basis but involve a large capital outlay. Second, unless breeder reactors are developed, which produce more fissionable material than they consume, nuclear fuel, too, will soon be in short supply. For these reasons, the generation of power from the atom will probably increase slowly and steadily to satisfy a growing share of our energy needs, but no dramatic crash program is likely. Progress will be, and rightly so, cautious and conservative, because many features are new and untried on a large scale.

LONG-TERM SOLUTIONS

It is remarkable how, in the past, major breakthroughs in science and technology have come just in the nick of time. We can only hope that another miracle happens in our own generation to avert a major energy crisis at the end of this century.

Solar energy, geothermal energy and the fusion process represent three diverse sources of energy which would be relatively pollution free and essentially inexhaustible. These are attributes which oil and coal never had. Unfortunately, none of these alternative technologies are practical at the present time; indeed each awaits a breakthrough in basic science and so we cannot be certain that they will ever work on a large scale. Solar power will never do much more than heat the bath water until an inexpensive and reliable method is developed to directly convert the sun's rays to electricity. Geothermal energy will not satisfy more than local needs until the technology is available to drill through the earth's crust and tap the enormous heat reserves in the center of the earth. Fusion is the biggest hope, but no one yet has harnessed the process in a usable way. We have a few years grace in which, hopefully, the breakthrough will come.

If all else fails and the breakthrough does not come, which to the optimist is unthinkable, we are stuck with the present type of nuclear power based on fission. To satisfy all of the world's power needs by this means would involve a staggering proliferation of nuclear reactors, with the attendant problems of uranium mining, waste disposal, the accumulation of risks, however small, of a catastrophic accident, as well as the ever present problem of the proliferation of nuclear weapons. The prospect is not appealing to even the most enthusiastic supporters of nuclear power. We cannot profitably look too far into the future; the breakthrough surely must come.

8
Some Risk-Free Benefits

THE PEACEFUL ATOM

The principal concern of this book is the use of radiation in medicine and in power production where there is a small but real risk to the human population to be balanced against the considerable benefits. In order to present a more complete and objective picture of the peaceful uses of the atom, it must be pointed out that there are a number of special applications of radiation which are useful and beneficial, but which do not produce any hazard or risk to the general public. A minute amount of radiation may be involved for the scientists and engineers engaged in the specialized applications, but the public is not exposed to any radiation—even to small amounts comparable to medicine and power production. These may be styled "risk-free" benefits and will be described very briefly in this chapter.

ACTIVATION ANALYSIS

When materials are bombarded by neutrons, some elements (but not all) are made radioactive. When this occurs, the properties of the induced radioactivity are characteristic of the particular element, or group of elements, involved. The properties that can be identified and measured are:

(a) The type of rays given off. These may be alpha, beta or gamma rays, or any combination of the three.
(b) The energy of the radiation emitted. This is the easiest property to measure with sensitive equipment.
(c) The rate at which the induced radioactivity decays. The "half-life," i.e., the time taken for the activity to decay or fade away to half of its initial value, has a unique value for a given isotope.

Observing one or more of these properties is the basis of *activation analysis*, which can be used to identify and measure small traces of materials too small to handle by conventional methods of chemical analysis.

The use of activation analysis has grown with the increased availability of suitable neutron sources. Nuclear reactors, large accelerators such as cyclotrons, or man-made nuclides such as californium-252 are the most commonly used sources of neutrons for this purpose. A few of the more significant applications of activation analysis will be briefly described to illustrate the impact that it has made on many aspects of life, which range from the academic to the eminently practical.

The work of the *archaeologist* has been greatly aided by sophisticated techniques in activation analysis. A silver coin or a fragment of pottery found during excavations may be bombarded with neutrons, so that trace elements contained in it will be temporarily made radioactive and identified by the pattern of rays which they give off. Not only does this technique allow a much more accurate and detailed analysis than would be possible by conventional chemical techniques, but it is totally nondestructive, inasmuch as the valuable object is not broken, damaged or harmed in any way. The detailed analysis, especially of trace elements, enables an archaeologist to estimate the period and probable place where objects were made. It is a virtually foolproof way to identify a fraud.

One of the most colorful and exciting applications of activation analysis, frowned upon rather by the academics, is in *forensic medicine* and *crime detection*. A hair, a broken piece of glass, or a fiber torn from clothing found at the scene of the crime, can be matched with those found on a suspect with little room for error. If a detailed analysis of two objects shows identical patterns of trace elements, it is virtually certain that they came from a common source. It was largely on the basis of activation analysis studies of a hair from Napoleon's head that the presence of arsenic was discovered, which indicates that the exiled emperor and general was poisoned.

An important research area in *biochemistry* is the recognition of elements which are essential for life, although they need only to be present in minute quantities. Activation analysis played an essential part in the recognition of selenium, and to a lesser extent of chromium and tin, as essential elements. Furthermore, now that all of the more obvious elements essential for life have been recognized, activation analysis is virtually the only technique that makes much sense in the future search for additional elements, because of its exquisite sensitivity.

Geochemistry is the study of the distribution of elements in the earth, and indeed in the universe. This subject received a great boost from the introduction of activation analysis. This technique was involved, for example, in identifying 70 percent of the elements in the moon rocks collected by United States astronauts on lunar missions. The significant advances made in geochemistry in recent years have led directly to a major new theory of the origin of the elements. This new theory, based on the idea that the stars are nuclear furnaces in which elements are forged, is able to accurately predict the peculiar distribution of elements in the universe. This development in geochemistry has little impact on the man in the street, but represents an important advance in scholarship.

By contrast, there are few individuals whose lives have not been touched by development in *semiconductor technology*. Miniature radios, television sets, electronic calculators, to name but a few, have all undergone a revolution in recent years. They are smaller, cheaper, more reliable and work better; most of these improvements reflect advances in semiconductors. It is not the purpose here to describe this technology, except to mention that semiconductors are based on high purity crystals of germanium, which must contain traces of appropriate impurities if they are to work satisfactorily. It is possible to make germanium crystals so pure that the total impurity content is less than one part in a million. And yet techniques of activation analysis can identify and measure the amounts of each impurity present; this could not be accomplished in any other way.

A useful and novel method of *mineral exploration* has been made possible by the manufacture of a new nuclide by the United States Atomic Energy Commission (or the United States Department of Energy, as it is now called). The new material, which is called californium-252, does not occur in nature anywhere on earth. It was first identified in the debris of the hydrogen bomb

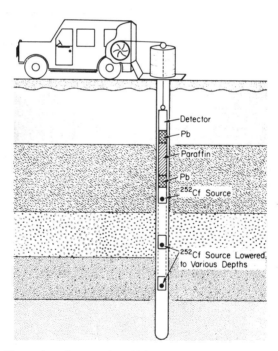

FIGURE 8.1. Illustrating the way in which a probe containing a californium-252 neutron source and detector may be lowered into a drill hole for activation analysis measurements as a method of mineral exploration.

exploded by the United States in the Eniwetok Island complex in the Pacific in November 1952. Small quantities can be made, though at considerable cost, in a few very special nuclear reactors, and the new nuclide is now available for use in medicine and industry. Californium-252 is unique because it can be fabricated into a small and portable source, which will give off an intense beam of neutrons for a period of many years (the half-life is 2.6 years). The source itself can be made as small as a thimble, but it must be stored in a large container, about the size of a 40-gallon oil drum, to protect personnel from the mixture of neutrons and gamma rays continually emitted. Housed in its protective shield, the source can be transported to the most remote corners of the earth to be used for mineral exploration. The technique, briefly, is as follows (see Figure 8.1). A small borehole is drilled deep in the ground, before the californium-252 source is lowered into place and left for a period of hours. The neutrons from the californium-252 radiate outwards from the source and cause the rocks and soil to become temporarily radioactive. The source is then removed and replaced by a sensitive radiation detector which picks up the rays emitted from the induced radioactivity in the ground. Since the nature and energy of these rays depends on the composition of the earth, the radiation measurements give a complete picture of the types of minerals present and their quantity. This technique depends on the availability of a *portable* source of neutrons, which was made possible by the production of californium-252. Until then neutrons were available only in sophisticated laboratories with access to a nuclear reactor or an accelerator such as a cyclotron.

PLANT BREEDING

Radiation is used to deliberately induce mutations in seeds of certain crops, particularly cereals. When seeds are irradiated, only a small number of viable mutations are produced, and most of these are useless. However, when a sufficiently large number of seeds are studied, a new strain is occasionally discovered which has desirable properties, such as improved yield and increased protein content.

This system of inducing mutations by radiation, and choosing mutants with more desirable properties, or which flourish in particular climatic conditions, is in effect a speeding up of the normal processes of natural selection, which are continually in operation in nature. It is potentially of great significance in countries where cereal crops form the staple diet. Indeed, the introduction into the developing countries in the late 1960s of new higher-yielding varieties of wheat and rice as a result of plant breeding gave hope to a food-hungry world of better things to come. The promise came, in large part, by the introduction of genes that imparted a short stiff stem to the plants, allowing them to absorb more nutrients in building heavier heads than traditional varieties. A complete package of improved variety, better disease and pest control, increased fertilizer rate, and better water management, resulted in yields which were doubled or tripled compared with those experienced in the past. By 1970, several countries, including Mexico, the Philippines and Turkey, which traditionally had been heavy importers of grain, had enough left over from their own needs to permit some exports. Several other countries, including India, Pakistan and Indonesia could begin to visualize prospects in the near future of becoming self-sufficient in good-grain production. This introduction of higher-yielding varieties became known as the *Green Revolution.* There were problems which surfaced early in the program, particularly the increased disease and pest hazard associated with a selected breed of cereal compared with the genetic diversity characteristic of wild types of grain. However, these were mostly overcome by further breeding programs. The real blow to the great promise of the Green Revolution came with the energy crisis in 1973. Suddenly, the all-important fertilizers and pesticides, so necessary to the successful performance of the miracle varieties, multiplied in cost and, moreover, were available in substantially less quantity than had been programmed in many of the developing countries. These shortcomings have, for the time being, taken considerable steam out of the revolution.

FIGURE 8.2. One of the miracle rice varieties, part of the green revolution, thriving in Indonesia. (Courtesy of the International Atomic Energy Agency Bulletin.)

LIVESTOCK PRODUCTION AND HEALTH

The International Atomic Energy Agency has promoted studies, using radioactive isotopes, to determine animal metabolism, aimed at increasing the efficiency of milk, butter and cheese production and at remedying deficiency diseases in livestock caused by a lack of trace elements in fodder.

Radiation has also been effective in controlling parasitic disease. The attenuation, or weakening, of larvae by irradiation has proved to be the sole method of preparing vaccines on a commercial scale for use against lungworm in cattle and sheep. Promising results have also been obtained by using this technique in the production of vaccines against such diseases as East Coast fever and sleeping sickness.

RADIATION STERILIZATION

Massive doses of radiation will kill bacteria, molds, yeasts and insects. It may be used, therefore, to sterilize medical supplies and to preserve food. For this purpose, high energy gamma rays emitted by a radioactive Cobalt-60 source, or x-rays and electrons generated in large electrical devices such as linear accelerators are of particular interest. The doses required to produce various chemical and biological effects are listed in Table 8.1

Table 8.1. Doses required for chemical and various biological applications of radiation

	Dose-Range	
Application	Megarads	Gray
Modification of the properties of polymers, e.g., polyethylene or PVC	5.0–25.0	50,000–25,000
Long-term ambient storage, e.g., meat	4.0–6.0	40,000–60,000
Inactivation of Bacillus anthracis spores in hair or fur	2.0–2.5	20,000–25,000
Sterilization of medical devices and pharmaceuticals	1.5–2.5	15,000–25,000
Sterilization of packaging materials for medical or food use	1.0–2.5	10,000–25,000
Rendering laboratory animal diets "pathogen-free"	1.0–2.5	10,000–25,000
Decontamination of food ingredients	0.7–1.0	7,000–10,000
Inactivation of Salmonella sp.	0.3–1.0	3,000–10,000
Reduction of micro-organisms in cosmetics	0.3–1.0	3,000–10,000
Extended storage of meat or fish (0–4°C)	0.2–0.5	2,000–5,000
Prolongation of fruit storage	0.2–0.5	2,000–5,000
Control of parasites	0.01–0.2	100–200
Control of insects	0.01–0.2	100–200
Inhibition of sprouting, e.g., in potatoes	0.01–0.02	100–200
Increase in mutation rate in seeds and plants	0.001–0.01	10–100

Sterilization of Medical Supplies

The principal commercial use of radiation for sterilization purposes relates to medical and surgical devices, which started in 1958 and has been expanding rapidly as more and more products and plastics that cannot be sterilized by heat or steam are found to tolerate the radiation doses needed for sterilization. Sutures, surgical dressings, pharmaceuticals and particularly plastic disposables such as syringes and blood transfusion kits, all sterilized by radiation, have revolutionized the work of the nurse, eliminated the chores of cleaning and reduced the incidence of accidental infection. Complex apparatus, such as heart-lung machines and kidney dialysis units, are frequently radiation sterilized before use. In medical research, too, vast quantities of disposable plastic culture vessels are used to grow human cells or microorganisms. These are prepackaged and sterilized by radiation and replace the daily task of washing and steam-sterilizing glassware.

Much of the pioneering work on the killing of bacteria by ionizing radiation was carried out at the Massachusetts Institute of Technology in the early 1940s with the specific aim of investigating the potential of radiation sterilization. For this work, the source of irradiation was an electron beam from a high energy van de Graaff accelerator. They found that bacterial spores were very resistant, requiring a dose of between 10 to 20,000 Gray (1 to 2 million rads) to kill them. In the late 1950s the use of radioactive Cobalt-60 sources for radiation sterilization of plastic medical products was explored in the United Kingdom and a dose of 25,000 Gray (2½ million rads) was found to be adequate.

There have been many arguments over the years, however, regarding the minimum safe dose necessary before a product can be regarded as sterile. The difference in philosophy comes in here—do you assume a product is contaminated with a reasonable level of common bacteria, or do you assume the "worst possible case," i.e., that the object to be sterilized is laden with the most resistant spores known to man? Depending on your philosophy, and the answer to this question, the radiation dose required would vary from about 20,000 to 45,000 Gray (2 to 4 *million* rads). Note how resistant to killing bacteria are—compared with a human for whom a few Gray (a few hundred rads) is lethal.

The source of radiation may be a large radioactive source of ^{60}Cobalt or an electrical device producing electrons. In the chemical area, where high doses are needed and a fast throughput, electron-producing accelerators are the usual choice because of their high dose-rate. Because electrons have a limited power to penetrate, accelerators are most suitable for in-line operations involving individual packages no more than an inch or so thick. Penetrating gamma rays from radioactive ^{60}Cobalt are particularly suitable for use in the sterilization of medical products where it is attractive to process medical

Table 8.2. Some possible applications of ionizing radiation to the treatment of food

Food	Main objective	Means of attaining the objectives	Dosage in millions of rads
Meat, poultry, fish and many other highly perishable foods	Safe long-term preservation without refrigerated storage	Destruction of spoilage organisms and any pathogens present, particularly *Cl. botulinum*	4.6
Meat, poultry, fish and many other highly perishable foods	Extension of refrigerated storage below 3°C	Reduction of population of microorganisms capable of growth at these temperatures	0.05–1.0
Frozen meat, poultry, eggs and other foods liable to contamination with pathogens	Prevention of food poisoning	Destruction of salmonellae	0.3–1.0
Meat and other foods carrying pathogenic parasites	Prevention of parasitic disease transmitted through food	Destruction of parasites such as *Trichinella spiralis* and *Taenia saginata*	0.01–0.03
Cereals, flour, fresh and dried fruit and other products liable to infestation	Prevention of loss of stored food or spread of pests	Killing or sexual sterilization of insects	0.01–0.05
Fruit and certain vegetables	Improvement of keeping properties	Reduction of population of molds and yeasts and/or in some instances delay of maturation	0.1–0.5
Tubers (for example, potatoes), bulbs (for example, onions), and other underground organs of plants	Extension of storage life	Inhibition of sprouting	0.005–0.015
Spices and other special food ingredients	To minimize contamination of food to which the ingredients are added	Reduction of population of microbes in the special ingredient	1.3

FIGURE 8.3. A modern potato irradiation plant at Hokkaido, Japan. (Courtesy of Dr. K. Umeda, Tokyo.)

FIGURE 8.4. This appetizing-looking meal has been prepared with irradiated foods. The beef steak was "preserved" by a dose of 50,000 Gray (5 million rads); the potatoes were treated after harvest with 100 Gray (10,000 rads) to inhibit sprouting. (Courtesy of the United States Army.)

devices in their final shipping carton or, for example, sacks of powder such as Kaolin in 50 lb. sacks. ^{60}Cobalt has a half-life of 5 years and is the most attractive gamma-emitter for industrial use. It is produced in a nuclear reactor—much of it in Canada. The first ever ^{60}Cobalt sterilization plant was built in England in 1960, and there are now over 90 in operation worldwide, with many more planned or contemplated.

Food Preservation

Sterilization of food by radiation is a big challenge at the present time. Complete sterility, needed for medical products, is rarely achievable with food because of changes of flavor that occur at high doses, to which the human palate is extremely sensitive. Lower doses, however, can kill insects

Table 8.3. Clearances granted for irradiated foods in various countries

A. *Clearances for public consumption (unrestricted amounts)*

Potatoes	Canada, Denmark, France, Israel, Italy, Japan, Netherlands, Philippines, Spain, Uruguay, USA, USSR
Onions	Canada, Israel, Italy, Thailand
Garlic	Italy
Wheat flour and/or wheat	Bulgaria, Canada, USA, USSR
Dried fruits	USA
Mushrooms	Netherlands
Dry fruit concentrates	USSR

B. *Clearances for experimental batches*

Potatoes	Bulgaria
Onions	Bulgaria, Hungary, Netherlands, USSR
Fresh fruits and vegetables	Bulgaria, USSR
Dried fruits and dry food concentrates	Bulgaria
Asparagus	Netherlands
Strawberries	Hungary, Netherlands
Cocoa beans	Netherlands
Spices and condiments	Netherlands
Prepared and semi-prepared meat products	USSR
Poultry	Canada, Netherlands, USSR
Fish	Canada
Shrimps	Netherlands
Foods for hospital patients	Germany (F.R.), Netherlands, UK

and retard spoilage due to bacteria molds or yeasts. Table 8.2, taken from an International Atomic Energy Agency publication, lists some possible applications of radiation to the treatment of food.

The potential advantages of food sterilization have been appreciated for over 25 years—indeed it has been said that food irradiation represents the most significant discovery in food processing since Nicholas Appert invented canning in 1810. It is attractive because it works without heating the product, it is effective within sealed containers as well as for bulk usage, and it does not leave chemical residue on the treated food. Nevertheless, up to the present time, food sterilization is not a large commercial operation comparable with sterilization of medical products. This may change in the future because of a recommendation made by the United Nations in 1981 that the process should be accepted by national health authorities up to a certain dose limit. The recommendations may find receptive ears because, faced with increasing concern over the safe use of chemicals such as ethylene oxide for inactivating microorganisms, or ethylene dibromide for insect control, health authorities are seeing the attraction of radiation. Another factor is the great increase of international trade in perishable food, since irradiation may be the answer to quarantine problems—as, for example, the import into Europe or the United States of exotic fruits from Africa—or for the treatment of food in the developing countries where so much spoilage occurs during storage.

Applications of Food Irradiation

Contamination of spices with bacteria molds and yeasts presents a serious problem since their addition to various foods causes spoilage. Irradiation at the recommended maximum dose of 10,000 Gray (1 million rads) kills most of the contaminants without a change in color or odor. Radiation sterilization is more expensive, but safer than the alternative which is ethylene oxide gas.

Radiation has been used to eliminate the dangerous pathogen *Salmonella* from frozen meat intended for dog food, and is used widely in Canada for frozen chickens and turkeys and in the Netherlands for frozen shellfish and frog's legs. A commercial scale irradiation of frozen shrimp was undertaken in Australia in 1979 to control a variety of pathogens. The treatments at 6,000–8,000 Gray (0.6 to 0.8 million rads) had no effect on taste or appearance, and the shrimp were sold through restaurants with no public reaction despite considerable publicity in the media. Any seafood eater who has been violently ill in the past from a favorite dish should welcome the security of knowing that the fish has received a sterilizing dose of radiation!

Radiation can be used to extend the storage of fish and meats which are refrigerated. Preservation of fruits and vegetables is also an attractive possibility. The high acidity in fruit is sufficient to prevent bacterial spoilage. The decay which can be controlled by irradiation is due to molds and yeasts, which require a relatively modest dose that does not cause changes of flavor or appearance. Mushrooms and strawberries have been treated successfully in this way in the Netherlands, which has been in the forefront of practical applications. Mangoes and papaya have been irradiated in large quantities in South Africa in acceptance trials, and can then be successfully shipped to Europe. It may one day to possible to enjoy the exotic tropical fruit in northern industrial areas thanks to radiation sterilization.

Insects and parasites succumb to comparatively low doses of radiation, but unlike the use of chemical agents, irradiation confers no control against reinfestation. Consequently, it is not applicable to grain handled and stored in bulk but is attractive commercially for products such as prepacked dried dates.

The inhibition by radiation of sprouting in potatoes, onions, garlic and shallots is well established as a way to improve the long term storage of these important root crops. The first regular commercial use of irradiation was for sprout control of potatoes during storage in Japan. A purposed built plant was commissioned at Hokkaido in 1974 and was handling 30,000 tons annually by 1978. Other countries, notably Hungary, Italy and Israel, have pilot plants in operation for onions and garlic as well as potatoes.

An interesting example of food sterilization is the "radiation sterlized" milk found in many supermarkets in continental Europe, which has a shelf-life of months and does not require refrigeration. In fact, milk itself cannot be sterilized with ionizing radiation since unacceptable off-flavors are produced by a dose as small as 100 Gy (10,000 rads), far too small a dose to kill the organisms that cause milk to sour. Sterilization is achieved by exposing the milk to ultraviolet (UV) light; to do so, the milk must flow in a thin film since UV is not very penetrating. Ionizing radiation, x- or γ rays, is then used to sterilize the plastic or cardboard container in which the milk is stored.

Safety of Irradiated Food

It has always been recognized that a clear demonstration of the safety of irradiated food for consumption must precede any commercial application of the process. Health authorities around the world have, justifiably, been very cautious, but having examined evidence of wholesomeness, have given permission for one or more irradiated food items to be consumed by the general public. Table 8.3 summarizes the present position. Among the irradiated commodities cleared in several countries are potatoes, onions, garlic, dried fruits, mushrooms, spices, and wheat.

Several United Nations Organizations including the World Health Organization (WHO) and the International Atomic Energy Agency (IAEA) collected data on animal studies and gave irradiated wheat, potatoes, papayas, strawberries, and chicken "unconditional acceptance." Onions, rice, fresh

cod and redfish received "conditional acceptance," meaning that more studies were needed. It was agreed that there is no toxicological hazard and no
general nutritional problems associated with food sterilization by radiation,
provided the guidelines are met. The approved types and sources of radiation
are not of sufficiently high energy to induce radioactivity in the food.

The Future

In most respécts, radiation sterilization would seem to be an ideal process;
it is safe, clean, efficient, easily controlled, capable of continuous operation
and measurable in terms of one single quantity—the radiation dose. In the
future, the biggest potential for radiation sterilization is in the developing
countries where so much spoilage occurs in stored food. In the Western
countries, refrigeration is so readily available that the need for new ways to
store food is not a pressing problem. The other major commercial application
is to overcome quarantine problems as well as spoilage as produce is shipped
from one country to another—particular for exotic fruits from Africa to
Europe and the United States.

INSECT CONTROL

The production of adequate food supplies for the large human population
on earth is critically dependent on controlling insects which feed naturally on
these foodstuffs. This control is routinely achieved by the use of powerful
chemical pesticides. Such control measures suffer the serious disadvantages
that most pesticides kill a wide range of insects, good as well as bad, and also
poison predators such as small mammals and birds. A much more sophisticated method of pest control is to capture some of the offending insects,
breed them in captivity until a large number is available and irradiate them
with gamma rays before releasing them to mix with the "wild" population.
Interbreeding between wild and irradiated insects produces no offspring, and
so the population gradually dies out. The enormous advantage of the sterile
insect technique is that only the offending species is wiped out, and in
addition birds and animal predators are not poisoned. (They may, of course,
go hungry!) In practice, it is not so easy to apply. Initially, the "wild"
population of insects must be reduced to a low level by using conventional
pesticides, so that the irradiated flies released are comparable in number. In
subsequent years, it is then possible to maintain a wild population close to
zero without any further use of chemical pesticides by a periodic release of
more sterilized flies.

A bonus advantage of the sterile insect technique, over and above environmental considerations, is that the ingredients for rearing insects in

FIGURE 8.5 (a). Method of producing large numbers of Mediterranean fruit fly larvae for insect control by the "irradiated male" method.

captivity are readily available in developing countries. This is not the case with insecticides, which usually must be purchased regularly and in vast quantities from the industrialized nations. Attempts are also being made to control the tse-tse fly by this method, but this insect is not so easy to breed in large numbers in the laboratory.

This technique has only been applied on a large scale in attempts to control the Mediterranean fruitfly, which each year ruins million of dollars worth of stone and citrus fruits. This fly has been the cause of a great deal of civic consternation and political turmoil in California during the past few years. Surviving on edible as well as ornamental fruits and berries, the insect has colonized a strip of land extending from Los Angeles to San Francisco. It has resisted a massive campaign in which helicopters sprayed malathion-laced bait and dropped millions of sterilized flies over vast tracts of the suburbs.

It thrives if hatched on a plant conducive to its nutrition; however, only certain fruits and vegetables can provide the nutrition needed for optimal development. It does best on peaches, apricots, mangoes, and guavas.

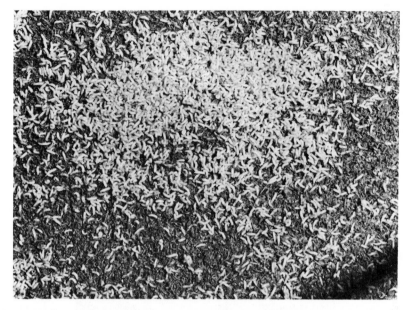

FIGURE 8.5 (b). Thousands of larvae being raised for insect control by the "irradiated male" method. (Courtesy of the International Atomic Energy Agency.)

The Medfly deserves recognition as a symbol of colonization. The first specimen noted by science was collected in 1817 on a ship in the Indian Ocean, where it probably hatched from fruit taken aboard in some African port. Within 150 years it has colonized most of the environment on earth in which it is capable of surviving. Its distribution ranges throughout the tropical and borderline regions of the world, with the curious exception of Southeast Asia, where it appears unable to compete with its rival, the oriental fruitfly. In California the fly is well established; it not only seems to be surviving in the face of the eradication plan, but to be expanding its range. The only note of optimism is that the chances are excellent that it cannot thrive in the San Joaquin Valley because the summers are too hot and dry, while the winters are too damp and cold.

The Mediterranean fruitfly, or Medfly as it is unaffectionately called, has sapphire eyes, a silver-speckled thorax and gold-spangled wings—not unlike hundreds of other varieties of fruitflys, but considerably more attractive than its distant and somewhat larger cousin, the housefly.

Even by insect standards, it is a fast breeder. In a lifetime that lasts only a few weeks, the female may produce up to 800 offspring. In fulfilling its maternal urge, the female inserts a needle-like ovipositor into a fruit or vegetable and spews 5 to 15 eggs into the puncture. It then moves on to another host to do the same again and again until the eggs are used up. The eggs hatch into larvae, which immediately begin serious eating. A strain of bacteria which live in close association with the Medfly are transmitted within the eggs, rot the fruit to make it more suitable as food for the larvae, and also cause the fruit to fall to the ground, making it easier for the larvae, when about eight days old, to emerge and burrow into the soil. Pupation takes only three or four days and soon after the adult fly rises from the soil ready to mate, and the whole cycle begins again.

The only practical time to kill the Medfly is during its adult stage, when it is out in the open. Most eradication programs exploiting the Medfly's taste for honeydew—not the fruit but a sweet liquid produced by aphids—have relied on a sweet bait laced with insecticide, usually malathion. In Israel, a country long plagued by the fly, farmers spray as a matter of course, but even so a sizeable proportion of the crop is lost each year. Malathion belongs to the chemical family of organophosphates. Some critics argue that it can cause mutations and cancer in the human population inevitably exposed. Certainly it produces mutations in cells in culture and chromosome aberrations in the blood cells of people inadvertently exposed.

A major technique used in California was the sterile fly release. By flooding the area with sterile male flies, fertile females would be most likely to mate

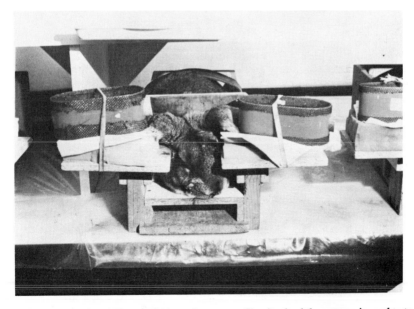

FIGURE 8.6. Early method used to raise tse-tse flies in the laboratory, in order to control this pest by the irradiated male techniques. The flies live on blood from the rabbit's ears. (Photograph by the International Atomic Agency, Vienna.)

with a sterile partner and produce no offspring, so the population cycle
would be broken. Millions of Medfly pupae bred in Peru, Hawaii and Mexico
were sterilized by radiation and raised to sexually active though sterile adult-
hood. Then they were tagged with fluorescent dust for later identification,
shipped to California and released. For several months, the program seemed
to succeed, but then fertile Medfly were found outside the original area. The
failure of the radiation sterilized flies in California was a serious setback.

FIGURE 8.7. It is now possible to grow tse-tse flies in artificial medium. Vast numbers
must be grown to control this pest by the irradiated male technique. (Photograph by
the International Atomic Energy Agency, Vienna.)

What went wrong? It is possible that the irradiated flies, brought from distant sites, were not accepted and could not compete equally for mates with the local population. But there is no direct evidence for this and early apparent success of the program argues against this as the principal explanation. It is more likely that either the estimates of the original Medfly population were too conservative, so that the number of sterile males released was inadequate, or that the fly had already spread beyond the borders of the target area before the program began.

Whatever the explanation, the failure of the program must be a serious blow to the future use and development of the technique. This would be most unfortunate since the alternative is constant spraying, which wipes out all insects indiscriminately and poses a hazard to human life—in place of the sophisticated targeted selective action of the sterile male method.

9
Risk versus Benefit

HOW SAFE IS SAFE ENOUGH?

Life is a very risky business. Sooner or later we all lose the game of chance and join the band of brothers for whom the bell tolls. The only question is whether we run our allotted span of three score years and ten before we die from "natural" causes, or whether we meet a premature and untimely end, victims of modern technology. Modern life in an industrialized society confronts us with a multitude of risks; some of these are obvious, like the risk of an automobile accident or an airplane crash—others are more subtle, remote and not immediate in their effect.

We tend to react to this problem of risk by making choices based on the magnitude of the risk as we see it, and the benefits to be gained from accepting the risk. Most of us, apparently, judge the speed and comfort of commercial air travel to be worth the risk of death that amounts to about one in a million for each flight. The convenience of driving our own automobile is considered worth much higher levels of risk! Sometimes our judgments are not especially rational. About 50 million Americans continue to smoke cigarettes despite the clear warning of risk to their health printed on each package, and the unequivocal association between smoking and lung cancer.

To a limited extent we are masters of our own destiny. We can refrain from smoking and thereby greatly reduce our chances of lung cancer. We can eat sensibly and exercise regularly to minimize the problems associated with obesity and hypertension. We can live in the countryside rather than the city, and by doing so escape a variety of hazards from murder and rape to air pollution. But our control is over a limited sector of life's hazards, unless we become hermits and withdraw from the mainstream of life. The hazards to which we are exposed are, for the most part, beyond our control.

RISKS AND THE STANDARD OF LIVING

In general, risks are associated with elevated standards of living. We drive motorcars, fly in airplanes, and heat our homes. All represent a step forward from primitive life in a cave, but each one carries with it a small but finite risk of a violent death as a result of human error or mechanical malfunction. The use of ionizing radiation falls into this category, too.

The uncommitted reader who has read this far must surely be impressed by the potential benefits of radiation in medicine and power production. One wonders how mankind could have managed without such a beneficent peaceful atom! On the other hand, enthusiasm for the unlimited use of radiation wanes rapidly after a brief account of the deformities in children, the risk of cancer and the genetic consequences to our descendants, which would be the inevitable consequence of excessive doses. There are two sides to this story.

As usual the extreme cases are easy to decide. There is little point in campaigning against nuclear power stations on the grounds that we may get cancer 20 years hence, if by so doing the lights go out and we freeze to death next winter. Where is the sense in refusing to have an x-ray picture taken because we fear genetic damage to future as yet unborn generations, when by declining medical care we do not survive to reproduce at all? These extreme cases are reminiscent of the slogans which were a common sight in the early seventies, such as "Make love, not war."

In constructing a risk-benefit equation there are only two sources of man-made radiation worth considering: medical x-rays and nuclear power stations. It barely makes sense to consider these two sources together because of the enormous difference in the dose levels involved. Medical uses of radiation give rise to doses which are 100 to a 1,000 times bigger than are ever expected from nuclear reactors used to generate power. The only thing that the two sources have in common is that radiation is involved; they are not comparable or competitive in any other way.

It makes more sense to compare the doses from the medical uses of x-rays with the radiation received from the natural background. At the same time it should always be kept in mind that the only sensible slogan is "no risk without a benefit." Once assured that this is the case, the possible deleterious effects of the radiation exposure resulting from medical x-rays may be assessed against the background level to which mankind has been exposed through countless generations.

The risks involved from the radiation levels associated with the generation of electricity from nuclear reactors should be compared with the alternative methods of generating power and with other comparable risks which we accept every day in an industrialized society; these in turn may be compared and contrasted with the hazards faced daily by mankind from natural disasters.

In general, the probability of an accident is related to the seriousness of its consequences. Disasters which involve death or injury to a large number of people are very unlikely to happen; on the other hand, accidents involving one or two people are *very* likely to happen, and a day never goes by without hundreds of instances occurring. The probability of a large-scale disaster involving the death of many thousands of people is dominated in industrialized societies by the remote chance of a dam bursting. Fortunately, this does not happen very often, but most adults will recall the death and devastation produced in northern Italy by an accident of this kind in the 1950s. At the other end of the scale, accidents involving a small number of people happen with high probability, and are dominated by automobile collisions, thousands of which occur every day. Intermediate between these two extremes are disasters involving the death of several hundred people, and this probability is dominated by air crashes.

The two principal sources of man-made radiation, medical x-rays and nuclear power stations, will now be discussed in terms of the risks and benefits involved.

POWER NEEDS AND RISKS

In a very real way, the standard of living depends directly on the amount of energy consumed.

In the years following World War II, power consumption increased dramatically in Western industrialized countries, but has plateaued to some extent in recent years as the price of oil escalated. The plateau, however, represents a very comfortable way of life!

By contrast, the Third World developing countries have barely tasted the good life, and will require enormous amounts of extra energy if their standard of living is to approach that of the United States and Western Europe. To be realistic, therefore, it is essential to plan for more power in the world as a whole, both in the short term and in the long run.

Any source of power, nuclear or conventional, involves a hazard to human beings and a threat to the environment. Those who adopt the attitude that any risk to man or to his environment is unacceptable are simply not living in the real world. Every facet of life in a technological society involves a risk. The ultimate choice as to how much additional power will be provided, and whether it should be generated from fossil fuels or from nuclear-powered stations, cannot be determined solely by the government, by power companies, by scientists, or by any other group. In the long term, society must decide what standard of living it wishes to enjoy, and must question whether this is compatible with maintaining the quality of the environment.

In the early days of the development of nuclear power, critics focused their attention almost exclusively on this form of power and never considered either the benefits or the risks of conventional types of power generators. This has all changed now because we are more aware of the energy crisis, and as a result the argument has taken on a larger frame of reference.

The choice is not nuclear power or nothing; the alternatives to nuclear power are coal and oil, which themselves are by no means innocent. The air

FIGURE 9.1. Fire in oil storage tanks within sight of the skyscrapers of New York City. The fire burned for days, pouring smoke and soot into the air. Such accidents are not infrequent; they often entail the loss of life of 2 or 3 workers, as well as a long term hazard from the pollutants. Yet they barely rate a mention in the local press! Compare this with Three Mile Island—where no life was lost.

pollution and damage to health caused by the burning of fossil fuels is now as important an issue as nuclear power. The shortages of gasoline and heating oil have sparked a new awareness in the industrial nations of the limited availability of fossil fuel resources. Not only is oil scarce, but its geographical distribution is nonuniform. The great industrial nations have grown up in Western Europe, the United States and Japan, while the Creator, in his infinite wisdom, saw fit to locate much of the oil elsewhere. As a result, there is a greater willingness now to look objectively at the overall advantages and risks of nuclear as well as other forms of power. A vital factor is that oil is essential for motor, rail and air transportation. At the present time, no practical alternatives are in view for these purposes. Consequently, there is even greater urgency to develop nonfossil fuel alternatives for central electricity power stations. It seems senseless to use the limited oil resources to generate electricity, when this can readily be done with nuclear reactors, reserving all of the oil for transportation, for which there are no alternatives at all.

ADVANTAGES AND DISADVANTAGES
OF NUCLEAR POWER

The benefit of nuclear power is that it is based on a new fuel that does not need to be mined in disruptive quantities, and whose flameless "fire" does not pollute the air with smoke and soot. The risks involve the chance of a catastrophic reactor accident plus the additional problem of waste disposal and routine releases into the environment of tiny amounts of radioactivity. In the long run, the principal worry concerning nuclear power may revolve around the export of reactors by Western industrialized nations resulting in increased membership of the "Nuclear Club." The benefit of coal is that it is abundant, at least in America, and that it avoids the risk of a catastrophic accident or the production of waste products that emit dangerous radiations; but it does involve other risks. Most obviously, it contributes in a big way to air pollution. Not only does a coal-burning power station emit clouds of soot and smoke, but the routine emission of radioactive material is *greater* than from a nuclear power plant generating a comparable quantity of power. Except in certain locations, coal is a deeply buried resource, and the mining of it is one of the most dangerous occupations on earth. In this century so far, over a hundred thousand miners in the United States alone have lost their lives digging coal out of the ground. Millions more have been injured or afflicted with black lung disease. Where coal beds come close to the earth's surface, it may be extracted by strip mining, but that practice so grossly lacerates the earth's surface that it has caused widespread popular reaction. Yet it is only through extensive strip mining that the coal industry can increase production enough to meet the growing demands for power production.

Nuclear power plants can be operated safely if their designs are carefully checked out, if high-quality control is exercised in their construction, and if their operation is subject to vigilant regulation at all times. The incident at Three Mile Island bears testimony to this; it is the worst accident in the history of commercial nuclear power production in the United States, and yet no member of the public was harmed. A further margin of safety could be gained by siting plants in remote areas, so that in the improbable event of an accident, the radiation risk to the population nearby is minimized. Some critics suggest that coal should replace uranium, but anyone who makes this suggestion has no concept of the vast amount of coal required for such replacement. By the end of the century, at least two billion tons would be needed annually, and this would involve an immense environmental assault, as well as an exorbitant social cost.

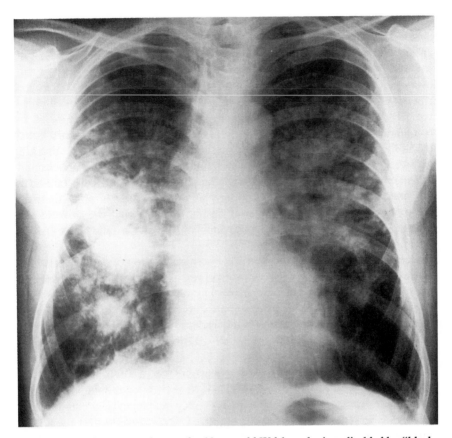

FIGURE 9.2. Chest x-ray picture of a 66-year-old Welsh coal miner disabled by "black lung disease" (pneumoconiosis). Note the opaque areas of lung which are non-functional. He uses two cylinders of oxygen per week and experiences discomfort in breathing in any but the erect sitting or standing position. (Courtesy of Professor Kenneth T. Evans, Welsh National School of Medicine, and Dr. David Hugh Evans, Cardiff.)

Given the alternatives, given their availability and their risks, nuclear power would appear to be an acceptable risk for the Western nations. It is the only practicable energy source in sight adequate to sustain the Western way of life and to promote the economy of Western nations. The important thing at the present time is to get an objective and realistic assessment of what those risks are.

It is a strange paradox that most of us are willing to accept one level of risk with one type of hazard, but very different levels with another type of hazard. For instance, automobile accidents represent a real killer in both Western Europe and the United States. Nevertheless, most people are willing to accept this very substantial risk as the price they pay for the convenience of using a motorcar. In the United States about 55,000 people are killed each year in automobile accidents; most of these are the occupants of motorcars, but some 10,000 are pedestrians walking on or near the highway who are run over, and about 500 are in their own homes, which are wrecked or destroyed when an automobile crashes into them. About 2,000 people are killed every year in the United States in air crashes; of these, about 200 are killed in commercial flights, 20 are innocent bystanders on the ground who are killed by falling debris, and the remainder are occupants of small private aircraft. Likewise, an airplane accident, in which hundreds of people die, shocks and horrifies people for a few days, but it has little impact on their long-term flying habits. Many more people are terrified of flying in an airplane than are frightened by driving in a motorcar, despite the fact that the drive to the airport is likely to be far more hazardous and involve a far greater risk of death, than a transcontinental flight.

Most of us take a very different view with respect to accidents over which we have control, or think we have control, versus those over which we feel we have no influence. The average person has very little control, in fact, over whether or not he is to be involved in an automobile accident, but we like to think that we do. On the other hand, radiation emanating from a nuclear reactor which is serving the community is regarded as inflicted on us by someone else.

There are other, and more tangible, reasons, too, for feeling a distrust of reactors. The fear of radiation goes back to the use of atomic weapons at Hiroshima and Nagasaki. In the minds of many people, nuclear energy is still equated in large degree with atomic bombs. The controversy over the testing of nuclear weapons above ground in the 1950s also played a large role in developing the overall public attitude towards radiation. Thus, the opposition to nuclear power is intertwined in the minds of most people with opposition to nuclear weapons.

MEDICAL USES OF RADIATION

The relative hazards to which we are exposed in life are summarized in Table 9.1. The risk associated with the use of radiation in medicine is conspicuously absent from this summary. It is not fashionable to speak of this, because it is usually supposed that medical procedures involve so much immediate benefit that the minute risks are more than justified.

The medical profession has a remarkable public image all over the world. We expect to be shortchanged by the automobile repairman, we stand still to be fleeced when our television needs service, but for the most part we trust our doctor! Despite the pressures that he, like anyone else, faces in terms of

Table 9.1. Risk of fatality by various causes*

Accident type	Individual chance per year
Motor accident	1 in 4,000
Falls	1 in 10,000
Fires and hot substances	1 in 25,000
Drowning	1 in 30,000
Firearms	1 in 100,000
Air travel	1 in 100,000
Falling objects	1 in 160,000
Electrocution	1 in 160,000
Lightning	1 in 1,200,000
Tornadoes	2 in 2,500,000
Hurricanes	1 in 2,500,000
All accidents	1 in 1,600
Nuclear reactor accidents	1 in 300,000,000

*Based on the Rasmussen study of Reactor Safety.

paying the rent and making a living, by and large we have confidence that our physician will not be influenced by economic pressure but will act in our best interests. This is perhaps less true than it was a generation ago, but we still expect our doctor to live by a moral code, and a set of ethics, superior to that of his fellow men.

This is not a result of advertising or brainwashing by professional medical associations, but rather a direct consequence of our love for life, our helplessness when faced with pain, suffering and death, and our intense desire to trust that someone can make us well again. For the most part, our trust is not misplaced. Radiology is no exception.

From the medical point of view, the small hazard to the patient from irradiation should be more than compensated for by the information obtained in the test, as a contribution to the diagnosis and treatment of his disease. If this is not the case, then there is no excuse for any x-rays to be taken. In the overwhelming majority of cases, the judgment of the physician is probably sound, but human nature being what it is, there are bound to be exceptions.

RADIATION THERAPY

It cannot be said that radiation therapy is used indiscriminately nowadays. X-rays are reserved strictly for the treatment of malignant diseases, and if a person already has a cancer from which he will die if not treated, then it is fatuous to worry about future hazards which may result from the radiation treatment, such as genetic effects in the future generation, or the remote possibility of a tumor caused by the radiation 20 to 30 years hence.

Few people who are treated for cancer subsequently become parents, and so worries about genetic consequences are not a big factor. That is not to say that cancer patients cannot and do not become parents—it happens regularly with no ill effects, but the numbers are small compared with the population as a whole, so their contribution to the genetic pool is negligible. At the same time, a patient who is being cured of his cancer by radiation has more pressing problems than to worry about the small possibility that the radiation itself will produce another tumor 20 years later.

If anything, it could be argued that we have erred on the side of being too cautious and too conservative in limiting the therapeutic use of x-rays so strictly to cancer. Some nonmalignant diseases do benefit from radiotherapy, and the benefit may be worth the risk. The present position is a reaction to the excesses of the past, where medical practitioners not trained as specialists in radiation therapy used x-rays to treat skin disorders in children—with good immediate results, but occasional later disasters. Such practices have quite rightly been eliminated, but it could be argued that it might be justified to use x-rays for the relief of pain in older patients suffering from complaints such as ankylosing spondylitis.

RADIODIAGNOSIS

The principal source of radiation to the public, dwarfing all other man-made sources, is the use of x-rays for diagnosis. On average, more than half of the people in the United States are x-rayed annually. This number varies from country to country as will be discussed later, but the figure is similar in most industrialized nations.

For the most part, x-ray procedures are a useful, and frequently vital, aid to diagnosis. But it would be naive to presume that all were necessary, and none superfluous. There are three principal causes of unnecessary radiation exposure of the public from medical x-rays. *First*, is the fear in which medical practitioners live of being accused of neglect or sued for malpractice. When faced with an accident case or an illness of uncertain origin, a cautious doctor may order a whole battery of tests, including x-rays, most of which are unlikely to be of any help, just to be certain that nothing serious has been overlooked, and to cover himself in the event that he is subsequently challenged or sued. No one is ever accused of "neglect" for taking a dozen superfluous x-ray films that were not needed, but one too few, if its absence leads to an incorrect diagnosis, can mean trouble for the doctor. This is sometimes known as "defensive medicine," practiced with one eye on the patient's lawyer.

Second, is the temptation, particularly in a private office or clinic, to order extra films that are not strictly necessary in order to generate additional income. When, for example, a new CAT scanner is purchased at a cost of close to a million dollars, it *must* be kept in constant use for 8 or 10 hours per day to justify its purchase and amortize its cost. Patients for whom the test is of marginal usefulness are likely to be signed up. It is, of course, difficult to prove this, but one suspects that it is true. At one point in the early years following the development of the EMI head scanner, there were more scanners in New York City alone than in the whole of the British Health Service—despite the fact that the machines were a British invention and made in Britain. One can only conclude that either hundreds of patients in New York were receiving head scans when they didn't really need them, or patients in Britain were sadly deprived of up-to-date medicine; the truth is probably somewhere between these extremes.

Third, it is difficult to ensure that all x-ray procedures are carried out by trained experts in radiology, since any doctor or dentist is licensed to operate an x-ray machine, although his training in this area may have been limited to one or two lectures in medical school. Furthermore, the field is changing so rapidly that keeping up-to-date is a major problem. Only in leading medical centers are the newest machines and the latest techniques employed, so that the maximum amount of diagnostic information is obtained with a minimum of radiation exposure. Through their various professional societies, radiologists are constantly trying to improve standards. Attempts are being made to introduce frequent self-evaluation examinations for physicians so that x-ray practices taught in the 1950s, when many of today's practicing doctors went to medical school, are replaced by newer and better techniques.

DOSES FROM MEDICAL RADIATION

Several detailed surveys have been made in the United States of the use of medical radiation, the most recent in 1970. A door-to-door survey of the population was made to discover the frequency and type of medical x-rays, following which measurements were made of the doses involved at a few representative hospitals. It was found that out of the total U.S. population of about 250 million, some 79 million had an x-ray in a hospital or doctor's office and, in addition, 59 million received dental x-rays. Some of course had both, and as a rule of thumb, approximately half to two-thirds the population in the United States are x-rayed each year for medical or dental purposes.

As explained in chapter 3, there are two principal long-term hazards of radiation. They are genetic anomalies, which affect future generations, and somatic effects such as leukemia and cancer, that will be of direct consequence to the individual exposed. For the individual, these risks appear small indeed. They become significant only when the nation is viewed as a whole. With more than half the population receiving x-rays every year, there is bound to be concern for the possible increase in genetic anomalies and possibly the production of malignant diseases, too.

Based on information collected in the survey described above, an estimate was made of the population dose relevant to genetic effects (The Genetically Significant Dose) and the dose relevant to the production of leukemia and cancer (The Somatically Significant Dose). They will be discussed in turn.

The Genetically Significant Dose

The Genetically Significant Dose (GSD) is estimated from the dose to the gonads (male and female sex organs) resulting from medical x-rays with due allowance for the reproductive potential of the people exposed. For example, if a woman who is past childbearing age receives x-rays, then it can obviously not have any genetic effect on future generations. These estimates are made by weighting the actual doses received by a factor to allow for the probable number of children the exposed individual may have. The GSD is the dose which, given to all members of the population, would result in the same genetic effects as the doses which are in fact received by some members of the population. The figure for the United States in 1970 was 200 microsieverts (20 millirem); this is compared in Table 9.2 with the corresponding doses from other man-made sources and also from natural background.

While this figure is the relevant one as far as genetic effects are concerned, it does not tell the whole story. Exposure of patients in the older age groups does not produce genetic effects because they will not have any more children—so they do not figure in the GSD estimates. However, older people, on average, have more illnesses and receive more medical care—presumably including x-ray pictures. Consequently these GSD figures conceal much of the medical radiation exposure of the population. The doses involved in a few common x-ray procedures are listed in Table 9.4.

A number of important and interesting points are at once apparent. First, medical x-rays in the United States, as in any industrialized country, dwarf all other sources of man-made radiation in contributing to the exposure of the human population. Nuclear power production is at the bottom of the list at present, less important even than color TV and luminous watches. As more and more nuclear electricity-generating stations are built, this figure will undoubtedly rise, but if all our energy needs were generated in this way, the radiation dose involved would never be more than a tiny fraction of that resulting from medical exposure.

Second, the total genetically significant dose to the human population from all man-made radiations is still much less than the average natural background radiation. It is the opinion of the various committees that have been convened to consider low dose radiation effects that man-made levels of radiation are unlikely to have any detectable genetic effect on the human population. This conclusion can be arrived at in two different ways. First, the annual genetically significant dose for medical radiation (200 microsieverts or 20 millirems) is lower than the estimated doubling dose, described in chapter 5 as the dose required to double the natural or spontaneous mutation rate (0.50 to 2.5 sievert or 50 to 250 rem) by a factor of several thousand. This appears to represent a large safety factor. The other line of argument goes as follows. Background radiation to the human population averages about 820 microsieverts (82 millirems) per year, but varies up to a value of 12 milli-sieverts (1,200 millirems) per year to significant populations of the world, as described in chapter 4, with no apparent ill effects or increase in congenital anomalies. It is concluded, therefore, that the addition of man-made radiation, in an amount which does not even double the average natural back-ground, is unlikely to result in a genetic disaster. To be more specific, one can live in a city such as New York, London, Tokyo or Toronto, where the background level is quite low, and have several x-ray films per year while receiving a radiation dose significantly less than the residents of the Rocky Mountain region receive from natural background alone. On this basis, it is unlikely that the present levels constitute a significant genetic hazard.

Estimating the Genetic Risk

It is possible to arrive at some very rough numerical estimate of the genetic impact of radiation from the information that is available. The following quantities are needed to estimate the number of genetic mutations per year resulting from the practice of diagnostic radiology in the United States:
Total U.S. population = 250 million
Approximate number of live births/year assuming a stable population and a life expectancy of 70 years $= \dfrac{250}{70}$ million

$$= 3.5 \text{ million}$$

Genetically Significant Dose per 30-year generation, due to medical radiation

$$= 30 \times 200$$
$$= 6000 \text{ microsieverts}$$
$$= 6 \text{ millisieverts}$$

Number of genetic mutations per million live births per Sievert of exposure in equilibrium from chapter 3 = 100,000
The number of genetic mutations produced per year in the United States by medical radiation is the product of the three quantities above, with the units adjusted to be compatible; $100{,}000 \times \dfrac{6}{1000} \times 3.5 = 2100$.

Of these, one-third, or 700, would be expected to be significant, and one-third trivial. These figures must be viewed against a natural background level of 350,000 genetic anomalies that spontaneously occur per year in the United States.

Comparison Between the United States and Great Britain

It is of interest to note at this point that a survey conducted in Great Britain in the late 1970s determined the annual GSD for medical x-rays to be about 118 microsieverts (11.8 millirem), compared with the figure of 200 microsieverts (20 millirem) for the United States in 1970. The lower figure in the U.K. presumably reflects the different pattern of health care delivery—a Nationalized Health Service compared with a free enterprise system. The socioeconomic implications of this difference are worth a moment to consider.

The number of films taken per million population varies widely from one country to another, with West Germany perhaps the highest. One would like to think that the practice of medicine, exemplified by the Hippocratic Oath, was motivated solely by the needs of the patient; but this is clearly not the case. As mentioned elsewhere, it has been estimated that in the United States one-third of all x-ray procedures are performed for reasons other than medical need. There are two principal motives. First, the need to generate income; in a private practice, when a great deal of capital has been invested in new equipment, the temptation to take extra films and add them to the bill is difficult to resist! Second, extra films are taken for what is frequently called defensive medicine, the practice of medicine with one eye on the patient's lawyer; unnecessary films are taken to be a safeguard in the event of a possible lawsuit. These same pressures are absent for the most part in Great Britain, but there, by contrast, it is observed that the number of radiological examinations varies with the cash allocation to a particular region of the National Health Service, implying that in some cases patients are denied an x-ray that the doctor might like to have as part of an economy drive.

In the U.K. between 1957 and 1977, the number of radiological examinations increased by 48 percent, yet over the same period the GSD had not changed significantly. This is due to the fact that the increased use of radiology is nonuniform and applied particularly to older patients, to the reduction in obstetric radiology and the real reduction in gonad doses achieved by better techniques. There was a marked increase in barium enemas between 1957 and 1977, reflecting the increased awareness of colon cancer and attempts to detect it early. There is a wide range of gonad doses from identical procedures carried out in different hospitals, and some places still have old and obsolete equipment. Gonad shields are used inconsistently. Direct shielding is particularly effective in the male and should always be used when it does not interfere with the quality and usefulness of the films. This is an area where the patient can contribute by questioning the doctors or technician performing the procedure. Clearly within limits, the use of x-rays in medical practice is influenced by social factors, which is evident from this comparison of GSD values between the United States and Great Britain, both sophisticated industrialized countries and both claiming to have "the best" health care system.

The number of genetic mutations resulting from the practice of diagnostic radiology in Great Britain can be estimated from the following pieces of basic data:

Total U.K. population = 60 million

Number of live births per year assuming a stable population = $\frac{60}{70}$ million

= 0.86 million

Genetically Significant Dose per 30 year generation due to medical radiation
= 30 × 118
= 3540 microsieverts
= 3.54 millisieverts

Number of genetic mutations per million live births per Sievert of exposure in equilibrium (from chapter 3) = 100,000.

The number of genetic mutations produced per year in Great Britain by medical radiation is the product of the three quantities above, with the units adjusted to be compatible, and amounts to:

$$100,000 \times \frac{3.54}{1000} \times 0.86 = 304$$

of which about 100 would be expected to be significant. This compares with about 86,000 spontaneous genetic mutations that occur in the U.K. annually from natural causes.

It should be noted here, as frequently repeated elsewhere, that these figures for genetic mutations produced by low doses of medical radiation are ESTIMATES, based on plausible assumptions and measured effects in animals at high doses, but they are no more than estimates and are certainly not facts or measured quantities in the human. It would be a statistical impossibility in any epidemiological survey to detect extra mutations as few as 100 against a natural background of 86,000!

Doses Relevant to the Induction of Leukemia and Cancer

In the 1970 survey, the bone marrow dose, i.e., the dose relevant to the induction of leukemia in the United States population from the use of medical radiation, was estimated to be about 925 microsieverts (92.5 millirems); this is almost five times higher than the genetically significant dose, because for cancer production, doses to older people are relevant, not just to those in their reproductive years, and of course the frequency of illness and the corresponding need for x-rays increases with age. This figure for the United States is compared in Table 9.3 with the corresponding doses from other man-made sources and from natural background. These same data are

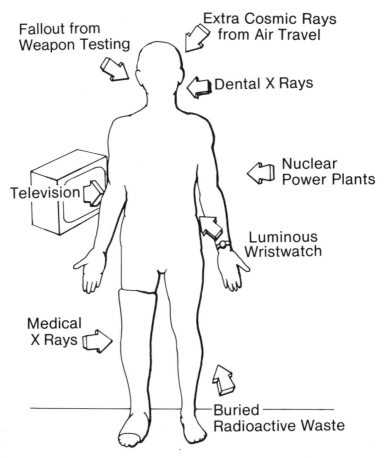

Fallout from
Weapon Testing

Extra Cosmic Rays
from Air Travel

Dental X Rays

Television

Nuclear
Power Plants

Luminous
Wristwatch

Medical
X Rays

Buried
Radioactive Waste

FIGURE 9.3. Illustrating the sources of man-made radiation to the U.S. population.
The dose equivalent is dominated by medical x-rays.

compared graphically in Figure 9.4, which shows that as far as the dose to
produce leukemia is concerned, medical radiation now accounts for more
than half of the total dose to the United States population. Other man-made
sources represent very thin slices of the total pie!

Estimating the Risk of Cancer and Leukemia

The impact of medical radiation on the induction of cancer and leukemia
in the United States can be estimated from the following quantities.
Total U.S. population = 250 million
Mean annual bone marrow dose, prorated over entire population, due to
medical radiation = 925 microsieverts (from Table 9.3).

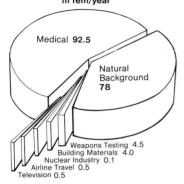

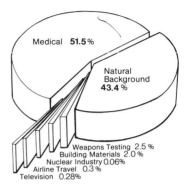

Total ≈ 180 m rem/year

FIGURE 9.4. *(Right side)* The radiation dose equivalent to the U.S. population from natural background compared with the various man-made sources of radiation. The doses quoted are to the bone marrow, and are pro-rated (i.e. averaged) over the entire population. *(Left side)* The doses in Panel A are here expressed as percentages of the total radiation dose to the U.S. population which is 1.8 millisievert/year (180 mrem/year). Medical radiation now accounts for slightly more than half of the total dose to the U.S. population; other sources of man-made radiation are relatively tiny. (Based on data in the BEIR III Report of the U.S. National Academy of Sciences.)

Risk estimate for radiation induced leukemia (from Table 3.1 in chapter 3) = 1.5 to 2.5 cases per thousand persons exposed per sievert.

The number of leukemias produced in the US by one year's practice of radiology is the product of the three quantities quoted above, with the units adjusted to be compatible and amounts to:

$$(1.5 \text{ to } 2.5) \times \frac{925}{1000} \times 250 \times 1000 = 347 \text{ to } 678 \text{ cases of leukemia.}$$

To the risk of leukemia must be added, of course, the risk of solid cancers. This is more difficult to estimate since doses from medical x-rays to organs other than the bone marrow are hard to come by. Based on the experience with the Japanese survivors, about three solid tumors might be expected for each leukemia; this would almost certainly be an overestimate since the Japanese were exposed total body, which is a very different situation to medical radiography. However, no other estimates are available, and on this basis the total number of cancers and leukemias in the United States resulting from one year's practice of leukemia would be about 2,000. These figures cannot be regarded as better than ball-park estimates, and it must be emphasized again that they are no more than estimates—based on plausible assumptions, but estimates nevertheless, not facts or measured quantities. It should also be noted that the current revision of the doses received by the Japanese A-bomb survivors may raise these cancer risks by a factor of 2 to 4.

The balance of risk versus benefit for medical diagnostic radiology, therefore, may now be summed up as follows: Each year, the practice of radiology benefits about 125 million people in the United States by providing a rapid diagnosis, but includes the *risk* of inducing a total of maybe 2,000 cancers, including 347–678 leukemias in addition to about 2,000 genetic mutations of which 700 may be serious.

While these figures cannot be regarded as better than crude estimates, which almost certainly overstate the risk, it is not difficult to draw the conclusion that the risk-benefit equation is weighted heavily in terms of benefit. For a patient who is sick and in need of medical care, the risk of carcinogenesis, which amounts to about 1 in 100,000, is a small price to pay for an accurate and speedy diagnosis, which is "safe" in that it does not, in general, involve invasive techniques, surgery or an anaesthetic. Nevertheless, one can never be content or complacent about the radiation doses received by the public in the course of medical diagnosis. Because the x-rays are given by devoted professionals, wearing impeccable white coats and working in a medical center, this does not mean that they are without risk. The medical profession and the public alike must insist that no exposure is justified, however small, unless there is an immediate and substantial benefit to the patient. "As low as reasonably achievable" must be the rule concerning radiation doses, coupled with the maxim, "no benefit, no exposure."

So long as there is a real and tangible immediate benefit to the individual or to society accruing from the radiation procedure, the small risk of carcinogenesis or future genetic mutations is clearly acceptable. This is almost always the case when the request for a radiological procedure is part of the diagnosis of an unidentified ailment. A different argument applies, however, to x-rays that are used for screening purposes, where a large number of perfectly healthy individuals will of necessity be irradiated needlessly in order to pick up the occasional case of ill health. The most striking examples are mammography, routine chest films, and routine dental x-rays. These need to be looked at rather carefully in order to ensure that the possible benefits outweigh the potential risk, and this will be attempted in the following sections.

DENTAL X-RAYS

Many dentists take a full set of x-ray films at each 6-month checkup on each of their regular patients. This may involve up to 12 x-ray exposures to get pictures of all of the teeth, each of which results in a dose of about 3 milligray (300 millirads) to the skin surface closest to the x-ray machine. In a

Table 9.2. Radiation of *GENETIC* consequence received by the population of the United States.*

Source	Average Annual Dose Millirems	Average Annual Dose Microsieverts
Natural background	82	820
Medical and dental x-rays	20	200
Nuclear medicine	2 to 4	20 to 40
Commercial nuclear power	< 1	< 10
Fallout from weapons testing	4 to 5	40 to 50
Consumer products (TV, building materials, luminous watches, smoke alarms, etc.)	4 to 5	40 to 50
Airline travel (cosmic rays)	< 0.5	< 5

*In the case of medical radiation, the figure quoted is the "genetically significant dose"—i.e., the dose to the sex organs weighted to allow for the reproductive potential of the individuals exposed. For all other sources, the dose is essentially whole-body and includes the sex organs.

Table 9.3. Radiation dose equivalents relevant to the possible induction of *LEUKEMIA*, i.e., the mean bone marrow dose, received by the population of the United States.*

Source	Average Annual Dose Millirems	Average Annual Dose Microsieverts
Natural background	78	780
Medical and dental x-rays and radiopharmaceuticals	92.5	925
Fallout	4 to 5	40 to 50
Nuclear industry	< 1	< 10
Consumer products	3 to 4	30 to 40
T.V.	0.5	< 5
Airline travel	0.5	< 5

*In the case of medical radiation, the doses are averaged and prorated over the entire population, although in fact only about half of the population is x-rayed each year.

Table 9.4. Average Doses for Selected X-Ray Examination

Examination	Dose to Sex Organs Male Millirem	Dose to Sex Organs Male Microsievert	Dose to Sex Organs Female Millirem	Dose to Sex Organs Female Microsievert
Skull	15	150	5	50
Chest	5	50	1	10
Shoulder	5	50	9	50
Upper gastrointestinal series	1	10	171	1710
Barium enema	175	1750	403	4030
Abdomen	97	970	221	2210
Lumbar spine	218	2180	721	721
Hip	600	6000	121	124

Source: Taken from gonad doses and genetically significant dose from "Diagnostic Radiology" as DHEW publication 76-8034, April 1976.

typical dental patient, adding up all of these exposures means that the entire oral cavity receives a radiation dose of about 3 rads (i.e., 3,000 millirads). Compare this figure with the doses from other medical procedures in Table 9.4 and from nonmedical sources such as natural background and nuclear power plants in Table 9.2 and 9.3. In defense of the dentist, it must be remembered that with adequate precautions he only irradiates a limited region of the body—not all of it!

Despite the sizeable doses involved in dental x-rays, a compelling case can be made for regular routine films in adults—perhaps not a full mouth series every six months but at least some films at somewhat longer intervals. Cavities which cannot be detected in any other way show up on films, and early detection may save a tooth, which is important because no man-made prosthesis can rival the real thing. It is not uncommon, too, for more serious problems, including cancer, to be spotted on routine films. There are tangible benefits to be weighed against the admittedly minimal risks.

It is harder to justify frequent and numerous x-ray films in young children. Finding a cavity in a "milk tooth" is no big deal, since the tooth will fall out and be replaced before adult life anyway. Some dental practitioners do not x-ray young children, or at least only at long intervals, and it is incumbent upon

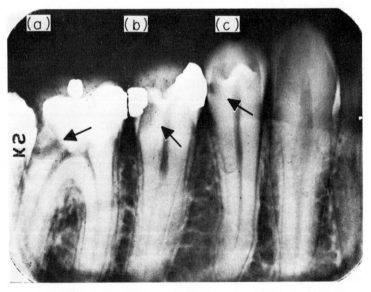

FIGURE 9.5. Illustrating the usefulness of dental x-rays to detect cavities that would not be found by direct examination. (a) A large cavity involving the nerve. The tooth will need root canal therapy, or extraction. (b) A small cavity under a filling only visible on a film. (c) A small cavity between teeth, only visible on a film. (Courtesy of Dr. Guy Webster-Towle.)

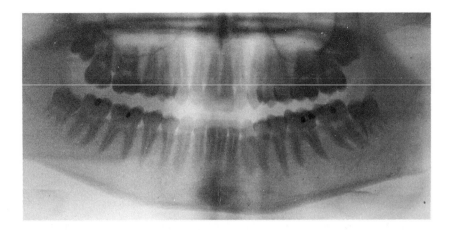

FIGURE 9.6. Panoramic dental x-ray picture. By using this modern device, called a panorex, an image of all of the teeth can be obtained with a single x-ray exposure. (Courtesy of Dr. Richard Salb and Miss Diane Hilal.)

those who believe otherwise to prove that the extra x-rays result in a real gain for the young patient. As a general rule, no amount of x-rays, however small, is justified unless there is a clear and tangible benefit to the patient.

What are the hazards? The Creator, mercifully, located the sex organs as remote as conveniently possible from the teeth, and if the dentist is reasonably careful the genetically significant dose should be vanishingly small, especially if a lead rubber apron is used to cover the body. The cancer risk too, is small but not negligible. The closest organ known to be particularly sensitive to radiation-induced cancer is the thyroid, and in terms of late risk this must be the principal concern (see Table 3.1, chapter 3). A full mouth series, carried out with good equipment and competent staff, results in a thyroid dose of about 200 microgray (20 millirads). Some dental schools, particularly conscious of radiation protection, have introduced a lead rubber "collar" to cover the neck in the region of the thyroid, but this does not help much since most of the dose to the thyroid when the teeth are x-rayed comes from radiation scattered within the body. This cannot be avoided.

A 1981 survey in Great Britain showed that, of a population of about 54 million, nearly 10 million receive dental x-rays each year. The genetic hazard was considered to be negligible, but it was estimated that there could be 3 deaths per year from cancer induced by the radiation involved. In the United States, no detailed survey has been carried out since 1970, at which time 59 million individuals had dental films. However, by 1983, the number of dental films sold by the leading manufacturer exceeded 700 million, implying that at

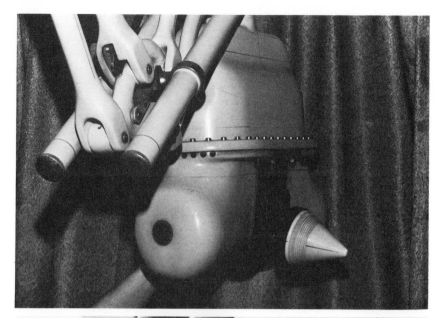

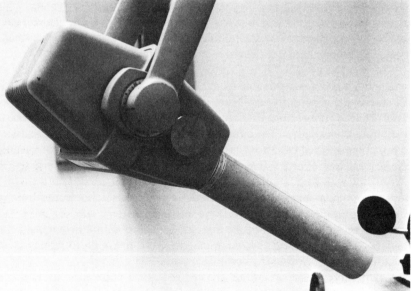

FIGURE 9.7. Pictures of old and new dental x-ray machines. (Top) An old-fashioned unit with a plastic cone. The x-ray beam is not well collimated and may lead to unnecessary irradiation of the patient while x-rays of the teeth are taken. (Bottom) Modern dental unit with a long lead-lined cone. This ensures that the area irradiated is the minimum needed to cover the film. The x-ray field size at the skin must not exceed three inches in diameter.

least 100 million (of a total population of 250 million) have dental films each year. While the genetic hazards are vanishingly small, this might result in 10 to 30 cancer deaths per year.

X-ray techniques used by dentists vary in quality perhaps more than any other medical use of radiation. Machines are often old and out-of-date, while modern techniques are slow to be adopted. Many of the dentists still practicing today were trained several decades ago, and are using equipment of comparable vintage. They are so busy filling cavities that there is no time to keep up with improvements in x-ray machines or techniques; besides, there is little stimulus to do so. The pressure to change and improve is most effectively applied by the patient. There are several things that you, as a patient, can do. Question whether films are really necessary, especially so many and so often; do not accept them as a routine procedure with small children. There is no excuse nowadays for any dentist to use an x-ray machine which is not fitted with a lead-lined collimated "cone" to limit the size of the x-ray beam; a plastic pointed cone is no longer acceptable. After all, the film placed in your mouth is little more than one inch square, and there is no point in having an x-ray beam which is any bigger. Why spray the whole head (and body for that matter) with x-rays to get a picture of a few teeth? In addition, insist that a lead rubber apron be placed over the body while films are taken. This prevents "scattered" x-rays (which bounce in all directions) from being absorbed in the body and especially the sex organs. If your dentist will not cooperate, change to one who will.

ROUTINE CHEST X-RAYS

Routine chest x-rays—those done in patients who are symptom-free, who appear normal on physical examination and who are not at high risk for lung or heart disease—are increasingly difficult to justify as medical costs soar and concern is expressed about radiation effects. Admittedly the chest x-ray involves less dose than almost any other procedure in radiology, but it is the most common medical test required as part of a health examination so that they account for nearly half of all medical and dental x-rays done annually in the United States. In 1980 this amounted to 75 million people exposed to radiation at a cost of nearly 2 billion dollars.

Chest x-rays ordered for individuals with a suspected health problem involves a positive benefit which clearly outweighs the minute risk involved. Those ordered for screening purposes do not. Millions of perfectly normal, healthy individuals are irradiated with no benefits, in order to detect the occasional abnormality. Tuberculosis is now rare in Western countries and can be screened for by using a skin test, so that routine chest x-rays to screen for this disease are not justified; they should only be used under special circumstances. Chest x-ray screening for lung cancer is more controversial. The issue at stake is whether earlier diagnosis on a routine chest x-ray improves a patient's survival anyway; it certainly does not justify the millions in which the test is negative. A sensible compromise might be to restrict yearly chest x-rays to heavy smokers and workers in high-risk occupations, such as uranium mining and the handling of asbestos. The same is probably true of the use of chest x-rays to detect heart disease; they should be restricted to patients whose histories or physical examination show that they are likely to have the disease.

For all of the reasons summarized above, the Federal Environmental Protection Agency in the United States recommended as long ago as 1976 that chest x-rays not be given as part of a pre-employment medical examination or as part of a periodic physical examination except among high risk patients. Consequently, in federal agencies, chest x-rays are not used routinely for pre-employment or periodic examinations, for tuberculosis screening programs, except in high risk groups, for hospital admissions of patients under 20 years of age and for prenatal examinations.

The stimulus to retain the routine chest x-ray comes from health care centers, for whom it is a substantial source of income, and from employers in the private sector who feel strongly in favor of the pre-employment chest x-ray for medico-legal and workers compensation reasons. The risk-benefit equation is not in favor of the individual in this case.

MAMMOGRAPHY

A highly specialized form of radiology that involves special emotional factors is mammography—the procedure whereby the female breast is x-rayed to detect the presence of cancer. This test came into the limelight in the mid-1970s when Betty Ford, the wife of the incumbent President of the United States, as well as Mrs. Nelson Rockefeller, were both found to have breast cancer.

Throughout the United States, doctors were besieged by millions of women demanding to be examined for breast cancer. Most of all, they wished to be reassured that they did not have the dreaded disease, but in case they did, they sought early diagnosis in the hope that it would ensure permanent cure. Things have calmed down now, but it is still an emotional issue.

There are, basically, three ways to detect cancer of the breast. First, the physical examination, whereby a lump or thickening of the breast tissue is detected by gently palpating or feeling with the fingers. This is not a very sensitive method of detecting a small lump, but at least it does no harm. Second, thermography, a technique in which the pattern of infra-red heat emanating from the body is recorded by a special instrument. Since this technique uses heat rays naturally coming from the patient, rather than passing x-rays or any other man-made radiation into or through the patient, it is totally harmless and without risk. Its success depends on the temperature of the skin and tissue over a tumor being higher than over the normal tissues. The value of thermography as a routine screening procedure has not yet been defined; it may prove to be very useful, but this has not yet been demonstrated.

The third technique is mammography, in which low voltage x-rays are passed through the breast and form a pattern or picture of any abnormal structure on the film which is placed beneath. Xeromammography is simply a specialized form of mammography utilizing a particular method of recording the data (Xerox electrostatic method) rather than the more conventional silver film method. Film mammography and Xeromammography, when performed by qualified radiologists, are similar in information content, even through the recording media are different. There is no question that mammography is effective and that tumors of the breast may be detected by this method earlier than they would otherwise be noticed. Against this, it must be admitted that an appreciable dose of radiation is involved. When the technique is used for screening millions of women, a proportion of those examined benefit by the early diagnosis of a malignant disease—but at the same time the majority are perfectly healthy normal young women who are unnecessarily irradiated with x-rays. Inevitably, therefore, the technique involves a risk of producing cancer as well as detecting it. The incidence of radiation-induced cancer from mammography will be very low, but when millions of women are involved, it will not be negligible. Consequently, screening techniques of this nature should only be used under ideal conditions, so that the risks are minimized. There are a number of factors to be considered.

(a) Most experts in the field recommend mammography as a routine screening procedure only for patients of 35 years or older. Younger women are only examined if there is already a definite suspicion of a lump that may be cancer. There are two reasons for this. First, the incidence of breast cancer is much lower in patients under the age of 35 or 40 than in patients beyond this age. Second, the imaging of breast cancer is much less satisfactory in the premenopausal breast than in the postmenopausal breast.

(b) The next question of concern is who should perform mammography, since considerable skill is involved in taking the films and even more in reading or interpreting them. It is clearly a job for a qualified specialist in radiology, and preferably someone who has made a special study of mammography. It is not a test that the average physician can perform on a part-time basis.

(c) The third relevant question concerns the equipment necessary for these procedures to be done properly. This is a very complex question and one for which a definitive answer cannot be given. Most would agree that the overriding factor is not the special equipment but rather the expertise of the radiologist making the examination. Mammography can be satisfactorily performed by the Xerox technique, or by film designed for mammography, and even by more conventional apparatus. The specialized equipment does allow the amount of radiation to be reduced, but is not vital to the success of the test.

This raises the question of the actual radiation dose involved in mammography. A series of films, taken by a skilled and competent radiologist, using adequately tested and calibrated equipment, involves a dose of about 0.01 Gray (1 rad) per film; since in a typical examination, two films for each breast are required, this amounts to a dose of 0.02 Gray (2 rads) to both breasts. This figure should be noted and compared with figures quoted for radiation from other sources. This is an appreciable dose of radiation; when millions of healthy young women are subjected to it, a few may possibly develop cancer from the radiation itself. Unless, therefore, mammography provides a substantial benefit to society by detecting many early cancers that would otherwise go unnoticed, there is no justification for using it.

In the mid-1970s, when millions of women in the United States demanded mammography, many of the tests were performed by doctors not trained in radiology, using unsuitable or obsolete equipment. As a result, a large number of women received doses far in excess of the optimal values quoted above, and even then the films produced were quite inadequate to detect the presence of cancer. Worst of all, many of the women tested were in their twenties, where satisfactory films could never be produced even by the most competent experts. The widespread use of this test under these conditions constituted nothing less than a scandal, and is a clear illustration of one circumstance where the uncontrolled use of radiation for medical purposes is vastly more dangerous to the public than all the testing of nuclear weapons and the generation of power from the atom that has even been suggested, much less performed.

The best advice to women contemplating mammography is this:

(a) If you are under 35 years of age, do not have the test unless there is some good and compelling reason to be suspicious.

(b) If over 35 or 40 years of age, regular mammography every few years is probably a good idea. Insist, however, that it be carried out by a qualified specialist in radiology. Choose a large hospital or university medical center for the test rather than the small private office, since a large center is more likely to have modern specialized equipment and a staff which is skilled in reading breast films.

THE 10-DAY RULE

The developing embryo or fetus is particularly susceptible to the effects of radiation. From the first to the sixth week following conception, the embryo undergoes "organogenesis." During this period, the various specialized organs of the body begin development, and radiation given at this time may cause one or more of these organs to grow in an abnormal way and result in a deformed or anomalous child. For this reason it is extremely important to avoid irradiating a pregnant woman, particularly in the early weeks after conception, which of course is just the time that a pregnancy may still be unsuspected. The only safe way is to insist that the so-called "10-day rule" be enforced, whereby a woman of reproductive age should only be subject to an x-ray examination during the 10 days immediately following the onset of a menstrual period of normal intensity, i.e., when she is quite sure that she is not pregnant.

This rule in practice can only be applied to "elective procedures," such as x-rays before changing employment or to monitor the course of a nonfatal disease; it cannot be applied to emergency situations where the life or health of the mother is at risk. Ideally this rule should govern the scheduling of x-ray departments, but many hospitals are so busy that they ignore it completely. Consequently, the logical person to enforce this rule is the person most likely to suffer grief and heartache if it is ignored—namely, the woman patient herself. Except in cases of emergency, where life or limb may be threatened by delay, insist that x-ray procedures involving the abdomen be delayed until after the start of a menstrual period. Even in an emergency, a woman patient should warn the doctor if there is a chance that she is pregnant, so that x-ray exposures may be minimized and every precaution taken.

In the later stages of pregnancy, the more fully developed fetus is less sensitive to x-ray damage, and in any case the pregnancy is so obvious that no doctor would ask for x-rays unless the health and the well-being of mother or baby were seriously endangered.

However much care is taken, there will always be occasional incidents when an early embryo is exposed to radiation as a result of diagnostic x-ray procedures, either as a result of an accident or an unavoidable emergency. Some workers believe that enough evidence exists to support the recommendation that 0.1 Gray (10 rads) to an embryo during the first six weeks of gestation is justification for an abortion, on the grounds that there is a significant chance of producing an anomalous child. This figure cannot, of course, be regarded as hard and fast. The recommendation must be interpreted with compassion and common sense. What makes the decision so difficult is that about 1 in 20 of all children born have an anomaly of some kind. Radiation in significant doses adds to the risk, but in a specific case it is never possible to attribute an abnormal child to radiation received, since defects produced by radiation are indistinguishable from those which occur naturally. When a doctor is faced with a patient who has received radiation to an early fetus, the easy and safe decision is to recommend an abortion to avoid all risks of an anomalous child. This may not, however, always be the compassionate decision. In a family where a child has been wanted for years, and where the chance of a future pregnancy is remote, a mother-to-be may choose to accept the small additional risk resulting from a dose of radiation. In other circumstances, a young family that already has several healthy children may prefer not to risk any extra chance of a defective child, but rather wait to add to the family at a later time. Clearly, the facts must be explained to the mother concerned, who must be fully involved in the final decision.

BALANCING THE RISKS

The hazards associated with exposure to ionizing radiations are small, but cannot be neglected. In seeking ways to put the risk into some perspective, it is instructive to make a comparison with the other hazards that we face continually in an industrialized society, such as driving a motorcar or smoking cigarettes. Three types of exposure to ionizing radiation will be considered.

(a) A chest radiograph which involves a very small radiation dose of approximately 150 microsieverts (15 millirems).

(b) A treatment dose of radioactive iodine (^{131}I) is used to control thyrotoxicosis. This is a treatment dose where the radiation is used to reduce the activity of the thyroid gland as opposed to a tiny dose used for diagnosis. This particular example is chosen because it is one of the very few uses of ionizing radiation in medicine for treatment purposes that does not involve cancer. It is described in chapter 5. The radioactive iodine concentrates in the thyroid and gives a very large local dose to this gland— thereby reducing the activity of the gland, as an alternative to having a surgeon cut part of it out. While the iodine is circulating in the blood, it delivers a dose to the whole body of about 150 millisievert (15 rems). This total-body dose plays no part in the treatment of the thyroid, but is an inevitable and unfortunate consequence of using radioactive iodine.

(c) The estimated average dose to the entire population of the United States from the 100 nuclear reactors that are projected to be in use for the generation of electricity by 1985 is 5 microsieverts (0.5 millirem).

The amounts of radiation corresponding to the three situations described above are very small; no deleterious effects to humans have ever been demonstrated from doses of this size. What is known is that large doses of hundreds of rads, as received by the survivors of Hiroshima and Nagasaki, do produce leukemia. To be cautious and conservative and to safeguard the public at large, it is assumed that the induction of leukemia is proportional to dose, and that a tiny chance of getting the disease is associated with even the lowest doses. Making risk estimates of this kind involves all of the approximations and uncertainties described in detail in chapter 3. However, if these uncertainties are accepted, it can be estimated that a chest radiography involves a chance of developing leukemia of approximately 2 in 10 million; the thyroid treatment involves a risk of approximately 3 in 10,000; while the presence of 100 nuclear reactors in the United States involves a possibility of death due to leukemia of 1 in every 100 million people exposed.

These risks can be compared with the risk involved in smoking cigarettes or driving a motorcar. First, cigarette smoking. Approximately 50,000 people per year die of lung cancer in the United States, most of whom are smokers; the average smoker consumes 20 cigarettes per day. If it is assumed that there is a linear relationship between the number of cigarettes smoked and the risk of cancer, then it may be calculated that one person dies of lung cancer for every 7.3 million cigarettes smoked. This is almost certainly not accurate, but it is just as valid as extrapolating data from atom bomb survivors, who received large doses, to patients receiving 150 microsieverts (15 millirems) from a chest x-ray. Turning now to automobile accidents; statistics are available which indicate that approximately 56,000 people per year are killed in the United States due to collisions or crashes. Approximately half of the entire population rides in motorcars (100 million), and the average driver covers 10,000 miles per year; consequently it may be calculated that there is one fatality on the road for every 18 million miles covered.

On the basis of these estimates, which are admittedly very crude, it is possible to compare the risk of death from radiation with the risk of death from an automobile accident or from smoking cigarettes. It turns out that one chest x-ray involves the same risk of death as the smoking of one and a half cigarettes, or driving on the highway for three and a half miles. The treatment of thyrotoxicosis with radioactive iodine involves the same risk of death as the smoking of about 2,200 cigarettes or driving about 5,000 miles on the highway. Incidentally, it should be remembered that this treatment avoids a surgical procedure in which approximately 1 in 1,000 patients die under the anesthetic. The radiation risk involved in the use of 100 nuclear reactors to generate electricity corresponds to a risk per year of smoking one-fifteenth of a cigarette or driving on the highway for approximately 100 yards.

It is evident from these considerations that while the risks of radiation are by no means negligible, they are small compared with the other hazards which we already face and accept in a Western industrialized society. The hazard in having a chest x-ray is equal to the risk involved in driving to the hospital by car, if you live three or four miles away, as most people do. A radioactive iodine treatment is much more risky and equals the hazard involved in 6 months average driving or 4 months cigarette smoking. The justification for its use is that it is less hazardous than the alternative method of treatment, which is surgery. The radiation risks from routine emissions from power reactors are extremely small, and negligible compared with other hazards—natural or man-made—which we face daily. For this reason, even the most vocal environmentalists no longer regard this as a serious problem with reactors, and now base their objection to nuclear power on the possibility, however remote, of a catastrophic accident. In this context, a serious accident involving the death of 100 or more people is conservatively estimated to be expected every 10,000 years! (see Table 7.2). This leaves the most serious objection to nuclear power worldwide as the proliferation of nuclear arms, which of course is not a problem within the Western industrialized nations which already have access to an abundance of such weapons, but does argue against providing nuclear reactors to developing countries.

CONCLUSION

The aim of this book has been to describe the properties of ionizing radiations and to discuss their use in an objective and impartial way. The reader must judge to what extent this goal of impartiality has been achieved. This book does not consciously or knowingly serve the purposes of any lobbying group for or against the use of radiation. In the minds of most people, radiation is still associated with atom bombs and the devastation of Hiroshima and Nagasaki. Radiation has always had a poor public image, which has been further tarnished over the years by very biased reporting in the popular press. There can be no question that radiation is potentially harmful; excessive doses have a devastating effect on living things. However, there is no direct, conclusive evidence of human disability, either in the form of cancer or genetic anomalies, arising as a consequence of doses of x- or γ-rays of about 0.01 Gray (1 rad) the entire dose range involved in medical radiography or in power production by nuclear reactors. Statements appearing in the press that a certain number of excess cancers will be produced are estimates, based maybe on plausible assumptions, but estimates nevertheless; they are not measured quantities or established facts. It is, however, prudent to take note of conservative estimates, since there may indeed be a real risk which is concealed by the statistical difficulties involved in documenting it.

A balanced view of radiation must also include an appreciation of the substantial benefits which result from harnessing the peaceful atom in both medicine and industry.

Radiation associated with the generation of nuclear power is very carefully regulated by the appropriate government agencies. Organized groups of environmentalists have been highly vocal and disruptive, but in the long run have been a force for good. Not only have they ensured that the existing regulations are enforced, but in several cases have succeeded in changing the regulations to make them more stringent.

The hazards associated with nuclear power are in fact very carefully controlled, though it must be admitted that the spokesman for the government agencies are conspicuously unsuccessful in convincing the public of this fact. It is a difficult proposition to explain to the man in the street that absolute guarantees of safety cannot be given in any area of human activity. This certainly applies to nuclear power, but it is by no means confined to this new method of producing energy. The problem is to explain simply, but convincingly, that the risk involved in operating nuclear power stations is very small and acceptable when compared with numerous other risks which are consciously or unconsciously accepted by the average citizen in leading an ordinary life—especially when weighed against the benefits and contrasted with the alternatives.

What is needed now is a pressure group, comparable to the environmentalists, to focus attention on medical and dental x-rays. At the present time doctors have carte blanche to use x-rays as they will; controls over the use of radiation in medicine are far less strict than those routinely exercised in the nuclear power industry. In general, the standards are high and improving all the time. Professional radiologists jealously guard their reputations, and try to minimize the misuse and abuse of radiation. When an x-ray procedure is needed to facilitate a diagnosis, the minute risks involved are dwarfed into insignificance by the immediate benefits. Unfortunately there are a few glaring exceptions to this rule, especially in countries like the United States where the practice of medicine is sold in the marketplace. It is difficult to curb the excessive taking of x-ray pictures when income is directly related to the number of films. It has been suggested that as many as a third of all x-ray pictures taken are not justified on medical grounds but are taken for financial considerations or as an insurance against malpractice suits. Mass screening procedures using x-rays, such as mammography or routine chest x-rays, are also open to debate. Their value has certainly not been proven, and they are particularly susceptible to abuse. If there is room for a crusade against radiation today, it must surely be in this area.

In general, the benefits which individuals enjoy as a result of medical x-rays are real and tangible. The balancing risks are small and hard to demonstrate. As long as this remains true, radiation will continue to be exploited in an increasing fashion, with the recognition that it is not only a powerful friend, but a potentially lethal enemy.

Index

About the Author

Dr. Eric Hall was born in a small coal-mining town in Wales. He was trained in physics at the University of London and in radiation biology at Oxford University, England, where he subsequently worked for a number of years. Since 1968 he has been Professor of Radiology at Columbia University in New York, where his time is divided between teaching and research. He is currently President of the Radiation Research Society, and his special interests include the effects of radiation on living organisms and the development of new techniques for the treatment of cancer. Research support for such interests is provided by grants from the United States Department of Energy and the National Cancer Institute. Dr. Hall's other publications include *Radiobiology for the Radiologist*.

Dr. Hall is married, has one son, and lives in Chappaqua, a village in Westchester County near New York City.